THE TITANS OF SATURN

CUMBRIA COUNTY LIBRARY

3 8003 03053 8771

3 0 AUG 2012		

629.455   GROEN, B. & HAMPDEN-TURNER, C.
THE TITANS OF SATURN

CUMBRIA LIBRARY SERVICES
COUNTY COUNCIL
This book is due to be returned on or before the last date above. It may be renewed by personal application, post or telephone, if not in demand.

C.L.18

THE TITANS OF SATURN

Leadership and performance lessons from the *Cassini-Huygens* mission

Bram Groen

and

Charles Hampden-Turner

CYAN

Copyright © 2005 Bram Groen and Charles Hampden-Turner

First published in 2005 by:

Marshall Cavendish Business
An imprint of Marshall Cavendish International (Asia) Private Limited
A member of Times Publishing Limited
Times Centre, 1 New Industrial Road
Singapore 536196
T: +65 6213 9300
F: +65 6285 4871
E: te@sg.marshallcavendish.com
Online bookstore: www.marshallcavendish.com/genref

and

Cyan Communications Limited
4.3 The Ziggurat
60-66 Saffron Hill
London EC1N 8QX
United Kingdom
www.cyanbooks.com

The right of Bram Groen and Charles Hampden-Turner to be identified as the authors of this work has been asserted by them in accordance with the Copyright, Designs and Patents Act 1988.

All rights reserved

No part of this publication may be reproduced, stored in a retrieval system or transmitted in any form or by any means including photocopying, electronic, mechanical, recording or otherwise, without the prior written permission of the rights holders, application for which must be made to the publisher.

A CIP record for this book is available from the British Library

ISBN 981 261 810 4 (Asia & ANZ)
ISBN 1–904879–41–1 (Rest of world)

Designed and typeset by Curran Publishing Services, Norwich, UK
Figures credit: David Lewis

Printed and bound in Singapore

CUMBRIA COUNTY LIBRARY	
H J	29/09/2005
629.455	£16.99

This work is dedicated to the late Earle Huckins III, who never had the chance to enjoy his dreams come to reality with the *Cassini-Huygens*' success at Saturn.

Earle, your innate ability to integrate opposing values is now forever symbolized in Saturn's orbit.

The authors and Earle's innumerable friends in the global space community

Contents

List of images	ix
List of figures	xi
List of boxes	xiii
Acknowledgments	xv
Foreword by Daniel Gautier, Tobias Owen, and Wing Ip	xvii

Introduction	**1**
1 Science versus politics: how the mission made it	**13**
The magnificent objective: the vitality of superordinate goals	14
Marshaling an international community	14
Apollo's dying dream	23
The funding crisis	27
The skepticism of Dan Goldin	30
The nest of paradoxes	31
2 The power of paradox: have values a logic of their own?	**33**
3 Errors and corrections: how to play seriously	**45**
Design and review	47
An error-correcting system	50
Simulating reality and playing to get serious	51
The self-repairing mission and the Pygmalion effect	53
Wide tolerances and built-in redundancies	55
Openness to supportive criticism	57
Scrutiny and trust	59
Qualities emerging from paradox	60
4 Competing to cooperate: how thousands of incomparable individuals fulfilled their common mission	**65**
In the wake of the Argonauts: a boat full of princes and captains	67
Furthering the ideal of Europe	69
Dangers and delights of diversity	70
Science as joint discovery	72
Heavenly sequences or "take your turn"	72
Twelve ways of sensing: interdisciplinary science	74
Cohesion, relaxation, and informality	76
Orchestrating the uniquely excellent	77
Qualities emerging from paradox	81
5 Engineering and science: lines and circles	**85**
Differences and dialogue between engineers and scientists	89

Europe as a partner with additional resources	91
The crisis of restructuring: losing the platforms	92
Setting up the trading system	96
Science gains over time	100
Qualities emerging from paradox	102

Taking a closer look: an introduction to the images section   **107**

6 Crisis and opportunity: how potential failure led to renewal   **113**

The environmental impact statement, launch nuclear safety approval, and public alarm	114
The probe's airflow problem	121
The Doppler shift on the probe relay	125
Qualities emerging from paradox	132

7 The equality of elites: leadership and culture   **135**

Unilateral authority dispersed	137
The resource provider	138
Film director and band leader	139
Defining the culture of the set	140
Integration and arbitration by team	141
Selecting for sociability and talent	143
Qualities emerging from paradox	144

8 Simplicity versus complexity: Goldin's cruel dilemma   **147**

An almost impossible job	148
"Faster, better, cheaper" – a polarized point of view	152
Command and control	154
Arch-skeptic and adversarialist	155
When qualities do not emerge from paradox	157

9 Lessons for Planet Earth   **159**

Lesson One: Mobilizing productive diversity	161
Lesson Two: Leadership shaping culture	166
Lesson Three: Increasing knowledge by integrating values	170
Lesson Four: Meaning as a prime motivator	172
Lesson Five: Learning through paradox	174

Epilogue: Tension, tears, and flow   **177**

Appendices

A The science instruments	185
B US–European collaboration in space science	193
C Text of letter from the director general of ESA to Al Gore, Vice President, United States of America	195

Notes	199
References and further reading	205
List of contributors	208
Index	209

Images

The images are located between pages 108 and 109

1. **International cooperation:** flags illustrating the international scope of the *Cassini-Huygens* program
2. **The interplanetary tour:** *Cassini's* four sister planet fly-bys
3. **Assembling the *Cassini* at Kennedy Space Center:** JPL technicians reposition and level the *Cassini* orbiter
4. **Attaching *Huygens* to *Cassini*:** the *Huygens* probe is installed onto the Cassini orbiter
5. **Lift off!** The *Cassini* spacecraft and *Huygens* probe begin their seven-year journey to the ringed planet
6. **Leaving Earth:** picture of the *Cassini* launch taken by Ken Sturgill of Marion, Virginia.
7. **Jupiter fly-by:** the solar system's largest planet, Jupiter, and its moon, Ganymede.
8. **Phoebe close-up:** first images from *Cassini-Huygens* of Saturn's moon Phoebe
9. **The magnificent rings:** *Cassini-Huygens* piercing the ring plane, capturing a stunning view of the dark portion of the rings
10. **Saturn's beauty:** the largest, most detailed, global natural color view of Saturn and its rings ever made
11. **The rings' kaleidoscope:** image showing the compositional variation within the rings
12. **The moon Iapetus**
13. **The moon Rhea**
14. **Titan's atmosphere:** a false-color composite of Titan'S hazy atmosphere
15. *Huygens'* **separation from *Cassini*:** an artist's conception
16. *Huygens*' **descent:** an artist's rendition of the probe reaching the surface of Titan.
17. **Titan's colorful haze:** the upper atmosphere shown in an ultraviolet image of Titan's night side limb, colorized to look like true color
18. **Pebbles on Titan:** a first image of the surface of Titan made on January 14, 2005
19. **Titan's surface:** the ground of Titan shown through two different infrared wavelengths
20. **Titan's Riviera?** An earthlike composite produced from images returned by Huygens during its successful landing

Figures

2.1	The challenge and skill difference	37
3.1	Error and correction	61
3.2	Accelerated learning	62
4.1	Competing to cooperate	81
4.2	Co-opetition	82
5.1	Engineering and science	104
5.2	Engineered discovery	105
6.1	Nightmare and awakening	122
6.2	Crisis and opportunity	133
6.3	Creative reorganization	134
7.1	Equality of elites	145
7.2	Egalitarian elitism	146

Boxes

The international team	4
The fascination of Saturn and Titan	8
Space missions to the large outer planets	16
History of discovery	22
Excerpts from a letter of 13 June 1994 from Jean-Marie Luton, Director General of ESA to the Honorable Albert Gore, Jr., Vice President of the United States	32
A homemade customer-driven scheduling solution: JPL's "RecDel"	63
Like a diamond in the sky	83
Engineers and scientists: some contrasts	90
The instrument with the black hand	98
Adjudication versus trading	99
The Stop *Cassini!* campaign	116
No "Tour de Saturn" without this Titan	124

Acknowledgments

When we set out to describe and study the *Cassini-Huygens* mission we never anticipated receiving so much support from so many people. In fact, almost 100 individuals made a direct contribution to our work, many through interviews we had with them wherever we were able to meet them. And a number of them will speak in this book. Yet jointly we know that they are merely the voice of literally thousands of people in the United States, Europe, and also in Asia who have at some time contributed to the success of the *Cassini-Huygens* mission. Some even dedicating their entire career to the mission. So we believe that this book should start with an acknowledgement of all these unnamed individuals. We decided to call them collectively "The Titans of Saturn."

The authors are indebted to the special contributions of Dick Spehalski, Wes Huntress, Jean-Pierre Lebreton, Dennis Matson, Enrico Flamini, Mark Dahl, Harley Thronson, David Southwood, Bob Mitchell, Toby Owen, and Daniel Gautier, without whom we would not have been able to write this book at all. In their respect for their late colleague and friend Earle Huckins, they let us into their literally borderless world of solar system exploration, helped us understand complex issues with patience, and reviewed thoroughly what we wrote about the facts. A special thanks to Dick Spehalski is in order, since he spent many hours with us at the Cape Canaveral condominium where much of this book was created, as well as to Harley Thronson and Mark Dahl of NASA who both never doubted the appropriateness of our mission. We are most grateful for both the tangible and intangible support from the National Aeronautics and Space Administration, the European Space Agency, and the Italian Space Agency in developing this book.

Then there are equally important people we have to acknowledge especially. There was Karen Chan, program secretary at the *Cassini* Program Office of the Jet Propulsion Laboratory in Pasadena, USA, and Ben Kroese, retired member of the *Huygens* team at ESTEC in Noordwijk, the Netherlands, who were both most helpful in connecting us to interviewees, enabling us to participate in key project meetings and finding important documentation. And there were Clare Miller, who typed innumerable chapter drafts, and PhD students Andrew Watson, Pi-Shen Seet, and Vesa Kangaslahti, who helped us research the mission, and planetary exploration and space policy in general, as well as in preparing the final details of getting quotation reviews and compiling literature references. The figures were drawn especially for this book by David Lewis. And, of course, our gratitude goes to our editor Susan Curran, who has that rare combination of skills of making what we think and say actually readable while not missing a detail or deadline.

We owe a special recognition to Fons Trompenaars, our partner at Trompenaars Hampden-Turner, and himself an accomplished author on the subject of cross-cultural competence and the inherently related logic of paradox. Fons brought the two of us, the authors, together, sensing that our combination of skills and experiences might integrate as well as they in fact have. As we developed this book, we synthesized our mutual capabilities in a way just like those who created this space mission did so well, merging state-of-the-art knowhow with practical solutions and orientations. It resulted in a unique point of view that we hope has rendered this book worthwhile for many who are – particularly globally – involved in industry, government, and academia.

We must particularly thank those close to us. Our spouses Sherry and Shelley helped us immensely in staying the course, editing our drafts, and enduring our concentration during all the long hours we had to spend making this a reality; and Bram's daughter Lisa and Charles' sons Michael and Hanbury cheered us on all the way, as did many other family members and personal friends. One cannot complete projects of this magnitude without such personal support, as many of the *Cassini-Huygens* families have also shown. In fact, that is how successful missions come about in the first place.

Last but not least, and to avoid the risk that we have not recognized others who we should have mentioned, we would like to express our gratitude to all those who participated in the development of this book (see list of contributors). It has not been possible to mention everyone in the text, but we want all to know that those who have been cited speak on behalf of all the NASA, JPL, ESA, ASI, academic and contractor team members we interviewed. We are honored that you spent the time to help create this.

Bram and Charles
May, 2005

Foreword

Living in the everyday world with its news of wars, natural disasters, and flamboyant personalities, it is sometimes hard to appreciate the magnitude of what our civilization has accomplished outside the tremulous canopy of air that gives us our sky. Future generations looking back on the period we are living through will surely consider the exploration of the solar system to be one of the most outstanding achievements of humankind. In our lifetimes, the planets, with their satellites, rings, and magnetospheres have been transformed from moving points of light in the night sky or dimly glimpsed disks and dots in our mightiest telescopes to familiar worlds with separate identities that are now known to hundreds of millions of schoolchildren all over the Earth.

We feel extremely privileged to have been able to play significant roles in helping parts of this extraordinary enterprise succeed. *Cassini-Huygens* is the largest robotic spacecraft launched thus far, and it is already sending back unimagined pictures and reams of fascinating data about our giant sister planet Saturn and its planet-like moon Titan. As our late colleague Hal Masursky remarked, "these spacecraft are our medieval cathedrals." This is what we built in our days to satisfy the endless human desire of connecting with the cosmos, through both our quest for understanding and simply our awe.

It is especially appropriate that this grand adventure had such a large international basis from its very beginning. It is a fine example of what humans can achieve when they put aside ethnic, religious, national, or disciplinary identities and simply work together toward a common goal. And what a goal we set! The two splendidly successful *Voyager* spacecraft had given us our first, clear close-up look at Saturn in 1980 and 1981. The richness of phenomena that they revealed made us eager to return with more capability to explore this miniature solar system in greater detail. That meant using an orbiter and a probe, something neither ESA nor NASA could do by itself at the time. Therefore the mission was widely supported by the American and European planetology communities.

This book by Bram Groen and Charles Hampden-Turner is very original compared with the usual writings about space missions. Instead of just describing the external aspects of the mission, it reveals in great detail how the mission was conceived, prepared, and accepted, while escaping the tentacles of some politicians in the United States seeking to cancel it in the mid-1990s. At that time, *Cassini-Huygens* was saved mainly due to the vigorous actions of ESA and the governments of the ESA member states.

The authors have interviewed an impressive number of participants in this unique adventure, so that many unknown aspects come to light when reading this book. And their logic of paradox puts a special light on this mission, that expands its significance from science and engineering accomplishments to an earthbound perspective in the field of management. It draws many conclusions about culture and values, and how to lead and reach top levels of performance, that suggest how our work was even more worthwhile, partly unbeknownst to us as we did so.

US citizens are already accustomed to the wonderful successes of NASA and the American industry in space missions. In Europe, the *Huygens* mission to Titan was, in 1988, the first ESA decision to build a vehicle designed to explore a planet, as we consider Titan to be. Europeans must realize not only that the ESA decision was audacious but that it would be very advantageous to Europe: the selection of instruments and investigators was guaranteed by both ESA and NASA as to be fully open to both European and American scientists. Once the selection was made, following a very competitive and intense process, it appeared that European (co-)investigators (over 130, no less) would be involved in all experiments aboard the *Cassini* spacecraft and the *Huygens* probe. Two principal investigators for the instruments on the orbiter and five of the six principal investigators for the probe's instruments are European; in addition, three of the nine interdisciplinary scientists are European. A fourth one, Jim Pollack, unfortunately passed away a few years ago, as have other important contributors to this mission such as Hamid Hassan, Harold Masursky, Fred Scarf, Gary Parker, and Earle Huckins. We appreciate the authors' wish to dedicate the book to the latter as a symbol for those who could not experience as humans the fruits of their magnificent work.

Huygens' entry into Titan's atmosphere demonstrated the prowess of the European space industry and has opened the doors for future joint ESA–NASA missions to the outer solar system, including we hope a return to Titan. And why not do so together with other space agencies such as those from Asia?

The success of the ESA–NASA cooperation in this mission can be seen in the results. We are still in the early stages of this mission, yet we are already admiring what these wonderful spacecraft are returning to us. Saturn seen by *Cassini*'s advanced cameras reveals a subtle beauty that *Voyager* could not quite capture. Those magnificent rings show much more structure now that we see them more clearly. The collection of instruments that explore the magnetosphere is revealing new phenomena on time and distance scales the *Voyager* science teams could not have hoped to study. And Titan and its atmosphere have proven to be just as strange and exotic as we had hoped. The *Huygens* probe was a triumphant success, with the images of Earth-like channels cut by rivers of liquid methane in Titan's icy, yet flammable landscape being one of the highlights of the entire mission.

This book tells the story of how this mission came to be. For those of us who initiated and now participate in it, the book provides a welcome opportunity for revisiting an important part of our lives. Our cooperative venture has proved not only scientifically successful but highly rewarding on a personal level, including making many new close friends. For almost every sentence in this book where you see our names attached to some action, you must remember that there were others, many others, who were working very hard to make this mission happen. Our early dreams were ultimately given substance by a huge team of scientists and engineers. It is why we agree with the authors that they are the true "Titans" of *Cassini-Huygens*' mission to Saturn.

Daniel Gautier
Emeritus Director of Research
Centre National de la Recherche Scientifique
Observatoire de Paris-Meudon, France

Tobias Owen
Professor of Astronomy
Institute for Astronomy
University of Hawaii, USA

Wing Ip
Professor of Astronomy
Institutes of Astronomy and Space Science
National Central University, Taiwan

April, 2005

Introduction

January 15, 1997
Cape Canaveral
It is 3 am Eastern Standard time, a warm Florida night, mosquitoes and alligators all around us. I am part of a gathering of several thousand scientists, engineers, officials, family, and friends who have all come to watch the launch of the Cassini-Huygens spacecraft. They are proud yet humbled by the sheer scale of what they have wrought between them, nothing less than a mission to Saturn and Titan. It is an excited yet anxious crowd. I hear voices in many languages, television stations from Europe and the United States, all mixed together in a modern-day Babel. Here are many unknown men and women, who have quietly worked on their dream – exploring Saturn and its moons – in search of understanding our solar system and Earth's origin. I realize that I may in fact be standing among the most creative and intelligent people on Earth, all come together in one place. What strikes me, watching this friendly crowd, is that a group of diverse people from all over the world is about to watch the product of a titanic team effort to advance science. What they have been able to do seems so hard to do in global business and politics: bridge their differences to realize a superordinate goal that dwarfs their own special interests. And now all are nervously hoping that the Titan IV-B/Centaur rocket will lift their dream into reality. Never mind that this specially modified version of the biggest rocket currently available on Earth has only been successfully tried once; its failure now is not an option.

The tension during the last minutes before launch is unbearable. The lights are off; the chatter is gone. Then ... five, four, three, two, one ... and lift-off. The sky lights up ... the thunder comes five seconds later. The spacecraft is leaving Earth ... slowly at first, then rapidly accelerating. The vapor trail is fantastic in the moonlit sky. Our reactions are tentative and muted at first; then we all burst into applause and jubilation, one big cheer in who knows how many languages. In just a few seconds all of the work of so many people could have failed. It did not; it worked, as everything in this program has done so far, despite enormous challenges. They have done it, all of them, together. Team Earth is on the way to Saturn on a seven-year journey. Never has a space mission so scientifically daring and significant thundered off into our solar system. And the people that made this work have all of a sudden become heroes to me, unsung heroes of a story that should be told.

So the story must be told – but what is the story? Is this simply another phase in the evolution of space exploration? When did it start and will it ever end, as this launch was just a milestone of even bigger things to come? Most of the information about such missions is on the Internet for many of us to read and study: the science, engineering, rockets used, and all the rest. But that was not really the point for me (Bram Groen, the Dutch co-author of this book who was present at the launch), much as the technology of solar system exploration fascinates me. No, my experience at this launch was the feeling that those present and those backing them had achieved something of immense significance. All their work up to that date had come together in one controlled blast of that Titan IV rocket. The paradox was that fallible people had achieved something superhuman, so that humanity could effectively "visit" a place almost a billion miles (1.5 billion kilometers) distant. If people can achieve as much across cultures and continents, what might we not accomplish? We knew instinctively that this was foremost a story about ourselves.

The first thought, of course, was that the US National Aeronautical and Space Administration (NASA) and its partners had done it again. After landing on the moon and visiting Mars and Jupiter, Saturn was next. It only takes money, technology, political will, and picking the best people. They have the best brainpower in science, engineering, and management, and therefore get things done well; end of story. But NASA's and other space agencies' histories have been chequered with several mishaps. Given the size of this mission, its vaunting ambition, and vast complexity, the chances of it functioning perfectly were slim indeed. Something unusual, even for space agencies, was happening here. Was this perhaps something you do not read about in most management books, something that transformed a level of aspiration most would regard as foolhardy into a triumph against the odds? What if, in fact, there was much more to this than the layperson can see?

To be sure, the same could be said about many other NASA missions, such as the ones to the moon or Project *Galileo* to Jupiter, as well as about the increasingly complex missions conceived by the Europeans to planets and comets. But what if NASA and all its partners had outdone even their own capabilities here? The message of the launch and this mission is that you can only find out if you explore, not knowing what you will find.

First, this author needed another, more intense emotional experience to make him really commit to finding out. Earle "Chuck" Huckins, his good friend and former *Cassini* program manager at NASA HQ, who had invited his friends to watch the launch, succumbed five years after the launch to the dreadful disease of amyotrophic lateral sclerosis (ALS). Chuck would never experience *Cassini-Huygens*' arrival and exploration at Saturn and Titan. We knew from him how this mission had given his life much of its direction and meaning, and that it was his legacy to those for whom he cared most: his family, friends, and colleagues. The Dutch have a word, *vergankelijkheid*, meaning the transitory nature of what is beautiful and magnificent. Upon his death, we resolved to try to capture this fleeting wonder and distil it in his memory.

The purpose of this book

Now, seven years after that incredibly powerful launch, we have come to see that this mission is indeed as successful as we assumed it would be at that time. The craft has arrived at Saturn, and the *Huygens* probe has successfully made its eagerly anticipated landing on Titan. Numerous publications and television documentaries have sung the praises of this mission.[1] Many more books will be written to explain the scientific discoveries made by the mission's instruments and the scientists behind them. It is evident that the *Cassini-Huygens* space mission to Saturn and its moons – a joint endeavor by the European Space Agency (ESA), NASA, and the Italian Space Agency, Agenzia Spaziale Italiana (ASI) – has indeed been an outstanding and improbable scientific and engineering success, generating more knowledge than any other space mission to date. Some predict it will generate more science than all prior missions combined.

At the time of this writing, all of this is occurring while the Shuttle is grounded, the Hubble space telescope may have to be abandoned, and space exploration has an ever-smaller share of US and European government budgets. The loss to date of three crews of astronauts has put US space policy in limbo. True, there are now new plans for human missions to the Moon and then to Mars, but how much of that is about the same politics as before? The scientific community feels even more slighted as this grandiose new idea has pushed funding for pure scientific discovery even further into the background.

Cassini-Huygens' extraordinary success is full of paradoxes. We travel to the far ends of our solar system to discover new truths about planet Earth. Saturn's message to Earth will be all about a system that is startlingly strange and new, and yet that also has much to teach us about our own humanity and how best to organize ourselves. It tells us how to learn, how to investigate, to discover and create. When human beings literally reach for the stars, in that very act they transform themselves.

With this book we aim to explore precisely that aspect of this human achievement. Ours is not a story of the impressive technical accomplishments that brought us so close to our large sister planet, or of the scientific discoveries made thus far and certainly to come. That we will leave to those far better qualified than us to explain. Nor do we intend to be comprehensive about or add new ideas to global space policy, or argue that human missions and other robotic explorations are not equally impressive on their own account. And rest assured that we will keep the science and engineering aspects of this story simple enough for all to grasp, as they frighten us as much as they may you. Our goal is more mundane and closer to Earth: to provide a new perspective and framework for how we, humans, can and in fact do achieve superlative performance, defining the "success" of a mission such as this one in another way while paying homage to those who brought success about. Drawing earthbound conclusions from this achievement can serve a broader purpose, which should not be lost in the excitement of the quite incredible engineering precision as well as the scientific discoveries made.

The international team

The *Cassini-Huygens* program is an international cooperative effort involving NASA, ESA, ASI, and numerous academic and industrial contributors. Through the mission, about 260 scientists from nineteen countries are gaining a better understanding of Saturn, its stunning rings, its magnetosphere, Titan, and the planet's other moons. The cost of the mission was approximately US$3.3 billion. This included US$1.422 billion for pre-launch development; US$710 million for mission operations; US$54 million for tracking; US$422 million for the launch vehicle; US$500 million from ESA and its seventeen member countries for the *Huygens* probe and science contributions; and US$160 million from ASI for instruments and spacecraft components. The United States contributed US$2.6 billion and the European partners US$660 million.

NASA

In the United States, the mission is managed by the Jet Propulsion Laboratory (JPL), Pasadena, California, on behalf of NASA's Office of Space Science. JPL is a division of the California Institute of Technology; it designed, developed, and assembled the *Cassini* orbiter. The US Department of Energy was responsible for the delivery of *Cassini*'s radioisotope thermoelectric generators and the *Huygens* heating units. NASA's Glenn Research Center was responsible for integration of the spacecraft with the launch vehicle and the design of mission-unique hardware and software modifications necessary for that integration. Under an agreement with NASA, the US Air Force was responsible for procuring the basic launch vehicle and conduct of launch operations. Lockheed Martin was the prime contractor to the USAF for the Titan IV-B program and its Centaur upper stage, to JPL for the spacecraft's propulsion module, and to the Department of Energy for the radioisotope thermoelectric generators.

ESA

Development of the *Huygens* Titan probe was managed by ESA's European Space Technology and Research Centre (ESTEC). The centre's prime contractor, Aerospatiale (now Alcatel), based in Cannes, France, designed the probe with equipment supplied by over 40 subcontractors across Europe as well as several from the United States. The *Huygens* probe was integrated in Ottobrun, Germany by DASA/Astrium, a subsidiary of EADS. The European Space Operations Centre in Darmstadt, Germany, managed the probe's flight and landing operations.

ESA is Europe's counterpart of NASA. It is an independent organization, although it maintains close ties with the European Union through an ESA/EC

Framework Agreement. The two organizations share a joint European strategy for space, and together are developing a European space policy.

ESA's seventeen Member States are Austria, Belgium, Denmark, Finland, France, Germany, Greece, Ireland, Italy, Luxembourg, the Netherlands, Norway, Portugal, Spain, Sweden, Switzerland, and the United Kingdom. In addition, Canada and Hungary participate in some projects under cooperation agreements.

ASI

The Italian Space Agency (ASI) provided *Cassini*'s high-gain antenna, the radar, much of the radio science system and elements of several of the science instruments. The Agency falls under the auspices of the Italian Ministry of University and Science and Technology Research. It has its headquarters in Rome. Italy was one of the first European countries to undertake space development projects, and in recent years has, together with NASA and other countries, developed a number of communications satellites building on Italy's unique capabilities in space communications technology, the reason ASI was involved not only through ESA, but also bilaterally with NASA for the radio science of *Cassini*. The Italian corporation Alenia Spazio was responsible for the development of ASI's contributions to *Cassini* and *Huygens*. In addition, Italy's Officine Galileo delivered the orbiter's star tracker, as well as science instrument components.

To get closer to this feat, we met and interviewed over 70 scientists, engineers, space agency officials, and others involved in the *Cassini-Huygens* program. We traveled to many European countries, to JPL in Pasadena, to the headquarters of the three space agencies involved, to several of the numerous subcontractors from US and European industry, and to the homes of those who had already retired. As we explored and shared their experiences and excitements, we began to appreciate the depth of their emotional commitment. We were struck by the passion apparent just inches beneath their professional reserve, for it was not unusual for them to weep as they told us their stories or spoke at press conferences. This was without doubt the most significant period of their lives, showing a confidence in the success of the mission born of tenacity. The human beings we met displayed an extraordinary respect for the expertise and the contributions of others, as well as deep satisfaction about their personal contributions, a satisfaction buttressed by much humility. Authority did not seem to be embodied in any person or persons, save in a formal sense. Rather it was "out there" among the planets, a force drawing them all on. Almost without exception these people were also passionate about wider issues, about how we are treating the environment, about the impending clash of civilizations, about the failure of other attempts at international cooperation, about short-sighted globalization: their perceptions were in no way narrow or specialized. It was as if they had seen Earth's fate in the stars and knew of vast extinctions.

With such attributes it was no wonder that Earle Huckins had been one of them.

We all live in a world of categories, boxes, straight lines, authority systems, bounded rationalities, *either–or* choices imposed by endemic scarcities, necessary compulsion, and a desire for predictable outcomes. Ours is an increasingly complex world – for global business, international politics, and so many other aspects of our lives – within which traditional linear thinking is rendered incomplete and single-principle imperialism is increasingly inappropriate. Yet on the leading edge of innovation – with a vast, unexplored universe before us – all these boundaries and traditions dissolve. We perch on the very edge of a vast sea of turbulence, with no prior experience to guide us or prepare us for so great a gamut of possibilities, so that our minds are cleansed of preconceptions and everything seems possible. If we need to be shaken loose of old habits of thinking, then Saturn and Titan are brilliantly qualified to transform us.

Saturn has been the wonder of our solar system for more than two millennia. On closer examination it has "rings of little ringlets," solar debris "shepherded" by tiny moons. The gas planet seethes with mighty storms of more than a year's duration. Saturn is the model for hundreds of "young stars" with luminous discs seen by the Hubble telescope. But the greatest excitement centers on Titan, the veiled moon/planet,[2] the only other body in our solar system with a nitrogen atmosphere like our own. It lurks within an impenetrable orange cloud, spews methane gas, and may have volcanoes of ammonia (the core ingredient of anti-freeze) and underground oceans.

So our purpose in this book is to point out that something very important has happened and is still happening, and that we ignore this rare feat at our peril. A group of over 250 scientists and up to 5,000 engineers and other professionals from nineteen nations have braved cross-cultural and multi-disciplinary misunderstandings to achieve an astonishing unity out of their wide diversity.

In a world threatening to come apart at the seams, this could be an invaluable model for a better world order. It is sometimes said that if humanity – if not only our politicians and global organizations – could discover a superordinate goal that dwarfs petty rivalries and pernicious dualisms, then the shared humanity of many cultures could be discovered even as that superordinate goal was achieved. "[It is] as if our gods, even the early gods, like the Titans, insist on an aspect of humanity in what seems like a godlike occurrence."[3] For us, the example set by the people of this mission has made them in our eyes the Titans of Saturn indeed.

Accelerated learning

By now it is axiomatic that we have entered the Information Age, where work becomes incredibly more challenging and complex. Yet we have been curiously slow to consider the ramifications of this reality, for example that information is self-generating and abundant; that "the more I give you, the more I have"; that dialogues between disciplines and cultures are mutually enriching; that learning, knowing, and creating are their own rewards.

No longer is it a question of performing tasks for which we have been professionally trained. Missions such as *Cassini-Huygens* are "ahead" of teaching institutions in this regard. No one in the world knows more about how to reach the Saturn system and which questions to ask. While knowledge is partly applied, it is also discovered and generated for the purposes of this mission. This book will inquire about how a whole project learns as it goes. If no one has done a particular thing before, from what source comes the authority to decide? Among a score of vital disciplines, whose is sovereign? How are conflicts resolved? How do they know they have got it right?

And how are these people motivated? It cannot be money, for their rewards are relatively meager. It cannot be "private enterprise." Funds come from three public agencies and more than a dozen governments. It cannot be "the American way"; nineteen different nations participated and a French, a Chinese, and an American scientist conceived this mission in the first place. Is this the product of "good old-fashioned competition"? Hardly. There was no competing mission for this destination. Was it due to a handful of heroic individualists? Perhaps, yet rarely have so many brilliant people had to rely so completely on each other, or accepted so much responsibility for the whole mission. It would be as true to speak of "heroic cooperation." What is clear is that the old clichés and folklore will not do.

Towards a logic of paradox

We have already touched on some peculiar paradoxes. We discover ourselves and our underlying humanity in shooting for the stars. Are we ever more alone than when we are creating and discovering? Yet we find ourselves in the fervent embrace of fellow enthusiasts. There is an amazing unity in the diversity of so many different disciplines and cultures. We need a superordinate goal, something "out of this world" to deepen our appreciation and wonder at the planet we inhabit and the amazing range of its human resources. In a strange, gaseous world of fuzzy boundaries, we also discover the limitations of our own categories of thought. Most people regard a paradox as a colorful anomaly, a temporary perplexity on the road to reason and order, a puzzle for science to solve. Could this be wrong? Is it possible to uncover *a logic of paradox* itself, with special relevance to situations of high complexity, novelty, turbulence, and discovery?[4] We take the view in this book that all those who generate new knowledge *are confronted with serial dilemmas and paradoxes*.

The Saturn mission encountered dilemma after dilemma, challenge after challenge, and surmounted them all. The idea that the *Cassini-Huygens* program was immaculately conceived and then generated without any hitch was an early casualty of our inquiries. Instead the mission appears to

The fascination of Saturn and Titan

The Saturn–Titan system, with its bright concentric rings and 31 known moons (with two more discovered recently by *Cassini*), has a powerful magnetosphere and gravitational pull. It is the second largest planet in our solar system after Jupiter. The planet's atmosphere above the clouds consists of approximately 94 percent hydrogen and 6 percent helium. The winds near Saturn's equator blow eastwards at 1,100 miles per hour (500 meters per second), making Saturn the windiest of the planets. It is very cold, with a temperature at Saturn's cloud tops of –218 °F (–139 °C). Its main ring system, spanning over 186,000 miles (300,000 kilometers), would barely fit in the space between Earth and our Moon. At the time of mission's arrival, the distance from Earth to Saturn was 934 linear million miles (1.5 billion kilometers), or ten times as far as the Sun is from the Earth); it will take some 84 minutes for light to cross that distance.

Why is this so exciting a planet to planetary scientists? Why travel a total of 2.2 billion miles (3.5 billion kilometers), since the route to Saturn is not linear but circular through our solar system? The answer in part is that this is a stellar laboratory that can provide answers to fundamental questions in physics, chemistry, and planetary evolution. Saturn has puzzled astronomers since the Middle Ages, when it would appear in various shapes in the sky. Depending upon the seasonal angle of observation, it looked like a triple planet, a planet with spikes, a teacup with two handles, or an elongated ellipse. We now know that its famous rings are made up of thousands of rock and ice particles, varying in size from cosmic dust to icebergs as big as houses. There are also multiple moons, which "shepherd" surrounding objects using gravitational pull. What we have essentially is "a ring of ringlets." The Hubble space telescope has shown us that there are huge storms on Saturn, and fierce cyclonic winds. There is a mysterious heat source on the planet, which radiates and cools, sending gases rising and falling again with great convective velocities. One theory is that helium migrates to the core and feeds a furnace of metallic hydrogen. Saturn is a low-density gas planet with few discernible boundaries between its atmosphere and its surface: a solar system in the process of formation. What we study there could be vital to the study of "young stars" and spiral disk galaxies, of which there are thousands. The planet and ring system serve as a physical model for the disc of gas and dust that surrounded the early Sun, and from which the planets formed. The success of searches for other planetary systems elsewhere in our galaxy partly depends upon how well we understand the early stages of the formation of planets.

Cassini's 76 orbits through the Saturnian system will include 45 Titan flybys, and the *Huygens* probe's descent into the moon's atmosphere on January 14, 2005 was designed to study its atmosphere and surface. A major goal of *Cassini-Huygens* is the unmasking of this large Saturnian moon. Titan is the only moon in the solar system that possesses a dense atmosphere (1.5 times denser

> than Earth's). It is particularly intriguing that this atmosphere is rich in organic material. Living organisms as we know them are composed of organic material. ("Organic" means only that the material is carbon-based, and does not necessarily imply any connection to living organisms.) In our solar system, only Earth and Titan have atmospheres rich in nitrogen. Earth's siblings in the inner solar system, Venus and Mars, possess carbon dioxide atmospheres, while Jupiter and Saturn resemble the Sun in their high content of hydrogen and helium. Hydrocarbons like the methane present on Titan may have been abundant on the young Earth. The importance of Titan in this connection is that it may preserve, in deep freeze, many of the chemical compounds that preceded life on Earth. Some scientists believe we will find that Titan more closely resembles the early Earth than Earth itself does today. Since all studies of life's origin are hampered by our ignorance about the chemical conditions on the young Earth, we need to know more about what starting material was present at the beginning of life on Earth. Titan may provide the answer to that question.

have been buffeted by a succession of enormous challenges, and even crises. Its real power lay in its resilience and versatility, and the creative improvization with which each hurdle was met. Its vicissitudes only seem to have made it stronger. Draconian cuts were ordered as superpower rivalry ebbed and it was no longer necessary to outspend and eclipse the Soviets. The cold war warriors had moved on. The triumph of Western economics meant privatizing anything that moved.

So far must this mission fly from the sun's energizing rays that plutonium fuel had to be used for powering the spacecraft's instrumentation. This triggered an almighty dispute with environmental lobbies. What if the spacecraft crashed back to Earth? What if the required Earth fly-by caused a collision? Patiently and painstakingly every possible disaster had to be imagined, described, weighed, and prevented. The program was also subject to a series of very stringent reviews. Review teams had to examine critically the work of fellow engineers and scientists. This even meant Americans evaluating Europeans, a process bordering on insult for many of the top scientists and engineers involved. Yet still the social fabric held.

But if this success was neither pre-planned nor crisis-free, it was not a coincidence either. "Fortune favors the prepared mind," and these people were well prepared. Yet some other dynamic than brilliant improvization was at work, and we feel challenged to explain what this was. We believe that outstanding people have always been fascinated by problems, paradoxes, and dilemmas, and have even chosen to confront these because of the challenge involved. We have given the art of reconciling paradoxes many popular names: genius, good judgment, insight, scientific breakthrough, discovery, creativity, innovation, problem solving, integrity – you name it.

We do not pretend to have found something not recognized or experienced before. Yet we do believe we have a new and coherent explanation of how such feats come about.

We apply our paradoxical logic not to the world of objects and materials. We have no quarrel with the scientific method. Rather, we apply this logic to values: the approach taken here is that superlative performance grows out of the special attributes of culture in an organization that has a clear mission. The elements that characterize a culture of stellar performance are reconciled and transcendent values. To grasp what this means we have to take a fresh look at values themselves. Values are not things, although we have given them names, such as "risk" or "courage," as if they were things. Our view, adopted from social anthropology, is that values are differences and hence always come in pairs, rather like red and green traffic lights that allow you to stop and go.

> This is the way the whole space enterprise should work: planet Earth moving into a new frontier. It should be an international enterprise and *Cassini-Huygens* is a good example. That is what I liked about this mission; it leads to better science and engineering. All [participants] bring something that is unique to the enterprise – approaches, cultures, knowledge. Such diversity always leads to better results.
>
> *Wes Huntress, former NASA associate administrator of space science*

So could Saturn be teaching Earth?

The problem up to now is that feats such as the mission to Saturn – and many other human and robotic space missions after *Apollo*'s moon landing – are recognized only by hindsight, and most of us simply share in the excitement of television images and newspaper stories, and then we forget. Those stories appear and disappear as mysteriously as they came. Many of us have had experiences, however fleeting, of intensely pleasurable involvement in challenges that yielded to our powers. It is not difficult to be nostalgic about such episodes or applaud success in others. What has eluded most of us up to now is a convincing description of the processes that culminate in outstanding performance. In this sense, the *Cassini-Huygens* mission to Saturn may well be a "laboratory" for the study of this phenomenon. Flashes of genius amid ashes of failure were not tolerable in this case. *Everyone* had to excel, and it is this synthesis of superlatives that we want to understand.

There are, of course, some quite extraordinary circumstances that gave rise to this mission. Rarely, if ever, in the history of civilization has such a wide spectrum of erudite persons wanted so desperately to know so much about a planetary system that has fascinated humankind for centuries. It was not done for profit; that requires a very different calculus of expected costs and gains. Rather, we did it because it was possible, and we yearned both to know and to realize our potential. That this knowledge and capacity can be turned to profit in the longer run we do not doubt, but the motive to perform superlatively lies beyond material gain.

Which is why the real import of *Titans of Saturn* is not just the story of this mission – impressive and outstanding although it turned out to be – but how we can meet the whole gamut of contemporary challenges with achievements of this kind. We must combat plagues that are almost certainly in part a consequence of malnutrition and unequal distribution, prevent global warming, conduct business in ways that exploit neither our physical nor our social environments. We must marry technology with humanity, overcome gross inequalities of power, and heal the wretched of the Earth.

For perhaps the biggest paradox of all is that you have to leave this planet and look back upon it to fully appreciate its fragility and uniqueness, a tiny oasis of life in the immensity of a cosmos, which is otherwise lifeless as far as we can see and wholly indifferent to our fate. It is when we band together to satiate our burning curiosities about things "out there" that we reach new pinnacles of excellence "in here." So ours is an exploration of the dynamics of superlative performance: how a space mission to Saturn could teach Earth to excel, speaking through our own Titans of Saturn.

1

Science versus politics: how the mission made it

It is difficult to say what is impossible, for the dream of yesterday is the hope of today and reality of tomorrow.

Robert Goddard (1882–1945)
American physicist, the father of modern rocketry

Even allowing for exaggeration by its advocates and admirers, *Cassini-Huygens* is delivering more science information than most or all of the space missions we have flown, from anywhere. There have been successful robotic missions previously: *Pioneer 10* and *11*, the *Mariner* and *Voyager* series, the Earth Observing System and the heroic *Galileo* among others. Yet *Cassini-Huygens* exceeds their combined outputs by multiples. Science was the space agencies' justification for this mission all along, so how has the scientific ideal been realized so spectacularly and so recently?

A popular management tool for assessing strategies is called SWOT: it stands for strengths, weaknesses, opportunities, and threats. Here we have two pairs of dilemmas or paradoxes. Can you somehow mobilize your *strengths* to shield your *weaknesses* and even turn them to advantage? Can you turn potential *threats* into *opportunities*? The early history of *Cassini-Huygens* shows how this feat was wrought. The ideal of planetary science has many weaknesses and faces many threats, especially from those who wish to use it for their own political purposes. Yet while politics can subvert science, it can also sustain and uphold it if the advocates of that science are wise and persistent enough. *Cassini-Huygens* fought its way through a political jungle and emerged from this more energized and determined than ever.

To understand the significance of *Cassini-Huygens* and the conclusions we draw later fully one needs to have some historical perspective, and the beginning of the book tells that story, while also introducing some of the key actors of Act One. "Fasten your seatbelts," as they say, because this was not a smooth ride. We will relay the story in five parts:

1   The magnificent objective: the vitality of superordinate goals.
2   Marshaling an international community of scientists.
3   *Apollo*'s dying dream.
4   The funding crisis.
5   The skepticism of Dan Goldin.

We begin by considering just how magnificent this superordinate goal was, and how it helped bring together a cohesive constituency. In the United States during the 1980s, the dream of *Apollo* was dying. The headlines of earlier years had left everyone with a hangover, while *Cassini-Huygens* was plunged into a funding crisis and had to confront the skepticism of NASA's new administrator, Dan Goldin. We shall finally reflect on the paradox of *science* and *politics*. Can politicians sustain the work of scientists?

The magnificent objective: the vitality of superordinate goals

Rarely in the history of space science has there been such a spectacular and absorbing subject of study. The Saturnian system is a miniature universe of its own, with millions of complex interactions within it. It not only attracts scores of different disciplines and observational instruments, but also draws them all *together*, because the answers to its mysteries lie somewhere between and among the phenomena.

It is a venerable principle of social science that the highest and most dedicated feats of individuality and cooperation occur when there is a superordinate goal, uniting otherwise very different people and interests.[1] All those concerned have something *varied* to contribute to the *unity* of the project. They need *others* to fulfill their *own* purposes. There are a thousand different facets to a single phenomenon. The superordinate goal is like the glass crystal hanging in the ballroom, in which the dancers can see their own and others' faces. Here lies a nest of problems to enthral science for a lifetime, a marvelously clear and simple goal, yet the gateway to a vast complexity. A rocket sent to a faraway place such as Saturn is a means to a thousand ends. Here was a priceless opportunity up for grabs.

Marshaling an international community

This mission was conceived, developed, and championed by planetary scientists of American, French, and Chinese origin, who organized a constituency of scientists in Europe and America, and influenced NASA and ESA with unified requests for support.[2] Tobias (Toby) Owen was an American specialist in planetary atmospheres.[3] In the early 1970s he had met Daniel Gautier, a French planetary scientist and recent recipient of the Legion of Honor and the Order of Merit.[4] Daniel was visiting the Goddard Institute of Space Studies in New York when the two got talking about the broadening of hydrogen lines by helium at Jupiter, and never looked back! Daniel, his wife, and his children met Toby's family, and all became firm friends. Daniel enjoyed America. Toby loved Europe. Both shared similar views of the world. It became a lifetime friendship, and moreover, the seed for achieving the super ordinate goal had been firmly planted. Toby told us:

> It sounds immodest but I like to think that this mission has benefited from our personal relationship. You have to overcome a lot of inertia and absorb a lot of defeats yet keep going. Agencies will not do that, only people who care enough. I thought naively that if we could do this, it would open up a whole new world of exploration by enthusiasts from different nations and countries. We'd all see how exciting it was. But that has not happened. Even today, we are doing less together, not more. Why two missions to Mercury, one from NASA, one from ESA? Why two agencies with missions to Mars? Why can't we explore the solar system together?

The idea to return to Saturn sprang naturally from the American *Voyager 2* project (see the box on space missions). It got to the vicinity of Titan and the view was dramatic but utterly mysterious. Titan, the "veiled planet," as many now have dubbed this moon that is in many ways more planet-like than moon-like, is wrapped in an icy orange shroud. Its thick, heavy atmosphere is mostly nitrogen and methane, and its persistence suggests that a continuous source of methane is issuing out of it, perhaps from an underground methane ocean. Going back to Saturn and Titan was the next logical step as many engineers and scientists on both sides of the Atlantic were now proposing.

Daniel contacted the National Centre for Space Exploration (CNES) in France, but CNES found that the project was far too expensive. He should find partners. Toby was receiving the same message from NASA, which was considering funding an orbiter around Saturn, but felt that the Titan probe idea would cost too much. Toby – and those US scientists who worked with him – wanted to use the *Galileo* probe that had already been designed and tested to explore Saturn, but NASA refused. To this day he wonders why. He suspects that the rivalry between JPL and NASA's Ames Research Center was simply too strong. Ames had built the *Galileo* probe, and JPL was having none of it. Toby then turned to Europe.

In the early 1980s Toby had been put in charge of the section of NASA's Solar System Exploration Committee that dealt at the time with future

Space missions to the large outer planets

While the space missions to the outer planets have played a minor role in NASA's scheme of things, *Cassini-Huygens* is by no means the first. *Pioneer 10* and *11* flew by Jupiter in December 1973 and April 1974, respectively. *Voyager 2*, launched in 1977, flew by Jupiter in 1979 and got within about 100,000 kilometers of Saturn in 1981, photographing the rings and four Saturn moons: Hyperion, Enceladus, Tethys, and Phoebe. It flew on to Uranus to discover a boiling ocean and winds of 450 miles (724 kilometers) per hour.

Voyager 1, launched after *Voyager 2*, returned some spectacular pictures of the Jovian moons. The eight active volcanoes on Io make it geologically the most active body yet seen in our solar system. The mission also discovered the moons of Thebe and Metis. *Voyager 1* went on to Saturn where it discovered five new moons and Saturn's C-ring. It passed only 2,500 miles (4,000 kilometers) from Titan, yet found its view blocked by dense orange clouds, 90 percent of which was nitrogen given off by organic compounds. The surface temperature was −292 °F (−180 °C). The craft then headed out of our solar system and is now the most distant human-made object in the universe at a distance of 8.7 billion miles (14 billion) kilometers from the Sun, and still going strong.

Galileo was launched in 1989, and arrived at Jupiter in July 1993. The craft's high-gain antenna failed on its outward journey; nonetheless, it returned no fewer than 14,000 images to Earth with the aid of a re-engineered low-gain antenna – a feat that deserves its own public recognition, not unlike this mission to Saturn.

Cassini-Huygens was launched in 1997 on a 2 billion mile (3.2 billion kilometer) journey to Saturn. No rocket existed to send it there directly, and four "gravity-assist" or "sling-shots" around various sister planets were needed to get there. The craft orbited Venus on April 26, 1998 and again on June 24, 1999. Then it came back to within 620 miles (1,000 kilometers) of Earth the following August, and around Jupiter on January 23, 2004. It reached Saturn on July 1, 2004, and at that point the most critical phase of the mission occurred – SOI, or Saturn Orbit Insertion – which took place without a hitch. The craft will stay at Saturn for at least four years and orbit the planet for science observations at least 76 times. The *Huygens* probe was released from the "mother ship" on Christmas Eve 2004, and splashed down into liquid ethane on Titan on January 14, 2005, after a parachute descent through the atmosphere that took two and a half hours.

exploration of the outer planets. Saturn was the obvious step after Jupiter, and the vision was to have both a Titan probe/radar mapper and a Saturn probe (as well as a mission to an asteroid and comet on a *Mariner Mark II* spacecraft, more about which later). While writing the report with his team, Toby was well aware that Daniel was also preparing a study on the same general subject for ESA in response to ESA's "Call for mission proposals" from Europe's scientific community.[5] The third member of the core team to be, Wing Ip, working at the time in Germany at the Max Planck Institute, had independently approached Daniel about this mission, and the two of them worked together to write the response to ESA.[6] Wing, a plasma scientist, is credited with having spurred on the European planetary science community in particular to throw their support behind the Saturn orbiter idea so as to further study the planet's magnetosphere. Wing had the bright idea to name the project *Cassini*, in honor of the great Italian scientist, who had come to France at the request of the King Louis XIV to be appointed as the first Director of the Observatoire de Paris.[7]

They all compared notes. The original European idea was that ESA would use a spare *Giotto* spacecraft to Saturn while the United States would contribute a Titan probe; as noted, the Americans thought of using a spare *Galileo* craft with perhaps a separate mission and probe to Titan. When Toby presented their views to NASA a few months later and its leadership told him there were insufficient funds for such "grandiose ideas," he suggested sharing the costs with ESA.

Although he had a solution to NASA's objection, the agency did not immediately like it, and would not explain why. Toby concluded that they did not want to depend on a "foreign" social structure so far from their experience, and feared loss of control. But Toby was stubborn. He had found equally stubborn friends and partners: a Frenchman, as well as a Chinese scientist working in and representing Germany. The three are very different, yet also so similar; Daniel comes close to what you have always believed Santa Claus would look like, bearded and ready for a bear hug, while Toby, tanned by the Hawaiian sun, and with his soft-spoken style, conveys the same friendliness and warmth so much exhibited by Wing. And all three are strong believers in international cooperation in science.

If the mission was now affordable, Toby persisted, why could his plan not be at least an option in his report? NASA agreed, and an option it became. In 1983, the Solar System Exploration Committee published its full report and suggested a possible collaborative international mission that would include a Saturn orbiter with a probe to Titan.

In the meantime, scientists had started organizing themselves on both sides of the Atlantic. In 1982, the National Academy of Sciences (NAS) in the United States formed a Joint Working Group with the European Science Foundation (ESF).[8] NASA and Congress will generally not consider a science proposal that does not have the backing of NAS. While Daniel and Wing had strong voices among their colleagues in ESF, Toby had to work through friends in the Academy. There were serious obstacles. Eugene Levy, the American chair of the joint committee, was a backer of a Mars surface rover and thus less in favor of the Saturn–Titan project.[9] Saturn was too far away, he argued. It would take seven years from launch to get there.

However, Fred Scarf and Hal Masursky, both close friends of Toby, argued strenuously for Saturn–Titan, and the three of them won over Hugo Fechtig, the chair of the European delegation on this joint committee.[10] The Joint Working Group ended up recommending several missions, one of which included a Titan probe and Saturn orbiter, both based on NASA's *Galileo* spacecraft designs. The idea was refined later that year with the suggestion that ESA should build a new lightweight probe instead.

Things were looking rosier when another setback occurred. NASA had just canceled its craft in the two-craft *Ulysses* mission to the Sun without bothering to inform its European partners.[11] ESA lost millions as a consequence, first hearing of the cancellation from the media. It was in no mood to embark on another grossly unequal relationship in which it had not been consulted.

France had an independent policy from America and the rest of ESA as far as the Soviet Union was concerned, and believed that joint space projects could thaw the cold war. The French aimed to collaborate with the Soviets on the *Vesta* project, which would investigate an asteroid and a comet. Perhaps that would pull them more closely into the international science community. Once again politics very nearly wrecked everything. Now space had to be emblematic of French independence! Of course, NASA had always acted as a prime contractor during its history. All other countries and organizations, ESA included, had thus far effectively been treated as subcontractors, as would happen again in the sixteen-nation International Space Station.

Toby credits Daniel Gautier's and Wing Ip's 1982 proposal to ESA with turning things around in Europe and eventually the United States. It was an important part of Daniel's, Wing's, and Toby's mutual respect that all three saw the "Big Picture." The Daniel–Wing draft did a masterful job of weaving together the potential contributions of many different sciences into one meaningful whole. A *whole* like Saturn is a miniature solar system in an early stage of development, a low-density "gas planet" providing vital clues on how solar systems form and solidify.

While Daniel and Wing were most persuasive, they arguably had a very fine subject to be persuasive about. A study of Saturn–Titan would provide the world with a crash course on whole-systems science, at a level of detail and sophistication never attempted before. Here were discoveries aplenty for specialists on icy satellites, on planet surfaces, on magnetospheres, solar winds and long storms, on atmospheres, plasma, cosmic dust, and even on the interior of Titan.

The mission to Saturn and its surrounding system was a laboratory in the sky, a clutch of lifetime opportunities to advance planetary science, bristling with many more scientific instruments than any previous mission. It would capture scores of variables present at a single point of time. Nothing this complex or ambitious had ever been attempted before. Daniel described it as "an exciting prospect, a vast spectrum of possibilities, and a beautiful opportunity for those who wanted to experiment." Nitrogen and several organic compounds that are present on Titan are also present in living organisms. This does not mean there is life on Titan. Life as we understand it requires liquid water, which is frozen on Titan. However, the carbon–nitrogen chemistry, like a primal soup in deep freeze, present on Titan may have been similar to what occurred on our own Earth some four billion years ago! Toby referred to it as "primordial ice cream."

What was so brilliant about Daniel's and Wing's draft was that it was a scientific paper with great political momentum. It organized its constituency on systems principles. Most of the scientific specialists and specialties mentioned in this paper were quickly won over to the necessity of a Saturn–Titan mission because they could clearly see their roles therein. The same was true of Toby's outer solar system team. Rarely, if ever, has a proposed mission attracted such a fierce body of adherents at such an early stage.

There can be no objection to politics, which is a part of science and brings together the people who illuminate reality from different angles, using different instruments. This was a kind of politics never achieved in the wrangles over *Apollo*, the Shuttle, the International Space Station, or the many unilateral or bilateral robotic missions in the United States and Europe. The early draft was turned into a fully fledged proposal with 27 co-signatories. And Toby's relentless efforts to unite the international science community and bring NASA and ESA into the fray, from the bottom up, mirrored the efforts of his European colleagues in forging this partnership.

> In these early days when we conceived the project, the standard was always "good science," and that transcended other often more directive solutions.
>
> *George Scoon, ESA project engineer during the feasibility phase*

For perhaps the first time in space exploration, a significant group of international scientists had spoken with one voice. NASA and ESA had always committed themselves to serve science, and now genuine scientists had unexpectedly seized the helm. It was difficult to object. The effort of the many scientists involved in these various studies and investigations eventually paid off: ESA and NASA agreed to conduct a joint feasibility study of a mission to Saturn. This was conducted by a transatlantic science team with in the lead ... you guessed it, Daniel, Toby, and Wing. Four scientists from Europe and four from the United States met repeatedly, expertly guided by the science-engineering duos Wes Huntress and Ron Draper of JPL, and Jean-Pierre Lebreton and George Scoon from the ESA side.[12] Ron and George and many other engineers from both sides were particularly important in this early phase, as they jointly led the effort of how to put engineering reality into the scientists' ideas. Both evidently had an engineering "feel" for what the scientists wanted. The team alternated their meetings between Europe and the United States to symbolize their intent of equal partnership.

All the different ideas were brought together, and the group ended up recommending in 1985 that NASA fund the launch vehicle and the Saturn orbiter, and that ESA fund and build the Titan probe.

> Even in those early days, I stressed with JPL and NASA that having two spacecraft – one to Titan with a Saturn fly-by, and one vice versa – was too complex and expensive. My view was that we should do one orbiter and develop it in the US. That's how it turned out to be later when all was said and done.
>
> Ron Draper, JPL Cassini *deputy project manager*

Even so, mobilizing this considerable weight of international expertise was not without its challenges. Many American scientists and engineers were initially worried that Europe did not have the technical capability to build the Titan probe. Toby was busy reassuring his fellow Americans that the Europeans were not, after all, rank amateurs. His greatest initial struggle was with JPL in Pasadena, the "arrogance" of which the Europeans deeply resented at that time. In those days he felt like a one-man diplomatic fire brigade. Yet his efforts paid off. By choosing to do the probe to Titan – the first probe ever tried by ESA – the Europeans had carved out a distinct role for which they would be responsible. Moreover it would, if successful, be the crowning achievement of the entire mission, an answer to the deepest mystery of the veiled planet.

While NASA had now given indications it wanted a mission to Saturn, and was investigating doing so with a *Mariner Mark II* spacecraft, ESA's Science Programme Committee had commissioned a Phase A study to define the specifications for the Titan probe. The Titan probe proposal was formally accepted by ESA's Space Science Advisory Committee in October 1988, after auditing six competing proposals each presented at a famous meeting in Bruges, Belgium, attended by over 300 scientists and engineers. Toby remembers feeling "very tense." The meeting took place in a cinema, which later that day would be showing the American film *Who Framed Roger Rabbit?* The director of science at ESA at the time was Roger Bonnet, whose opinion of American policy in space was not too favorable after the *Ulysses* experience. Toby felt the movie title was most inauspicious.

But Toby, Wing, and Daniel had been persuasive, not least as a result of the efforts of the NASA–ESA feasibility study team, as well as in particular the involvement of another French scientist, Michel Blanc. The "*Cassini* project" won backing over the *Vesta* asteroid and comet mission by eleven votes to two, but it was after heated debate while several other top scientists made a strong case for *Cassini*.

Until this moment the probe had not been named, although Toby remembers the name "Huygens" – the seventeenth-century Dutch astronomer who had first definitively identified Saturn's ring system – being subsequently suggested by the Swiss delegation on the Science Programme Committee. According to Roger Bonnet, he thought of the name in the midst of the Bruges meeting, checked it with a neighbor on the platform, and then announced it to the whole meeting, amid widespread applause and unanimity. Again it is likely that the same idea occurred to several people simultaneously, so obvious was the invocation of the names of these pioneers. Carl Sagan may have been the first to propose using the Huygens name several years earlier, in a letter to Daniel and others. Cassini and Huygens were both seventeenth-century giants in the field of astronomy, and the first astronomers to have discovered the planet's rings and moons (see the box on the history of discovery). The dual name was clearly intended to remind Americans of the historical roots of their own knowledge, while solidifying European unity around the early contributions of Italy, France, and the Netherlands.

That Bruges meeting accomplished much more. It not only gave the go-ahead to the *Huygens* probe, thereby anticipating that Congress would approve NASA's request for the orbiter, but also stipulated that in their negotiations with NASA both European and US scientists would need to be invited to share each other's instruments aboard both *Cassini* and the *Huygens* probe. As it turned out, European scientists participated in all twelve of the orbiter's instruments, while American scientists were involved with the six instruments used on the *Huygens* probe. What this arrangement meant was that while the Europeans had a distinct piece of the project, arguably the most exciting, there would also be the fullest integration of US and European scientists on all teams, built around shared involvement with each instrument. A sense of equality of Americans and Europeans had been designed into this one mission in a way never achieved before or since, not even in the International Space Station program, in which a number of European countries and Japan play major roles. Given that Americans were putting up 75 percent of the total costs, this was an equality of scientific expertise, not of economic power. In the days of *Apollo* and the cold war, such balance would have been inconceivable; limited access to American space projects was a diplomatic bargaining chip.

> Even in the cold war we cooperated with the Soviets, a perfect message of international cooperation and peace. It starts with being equal partners – a colonizing approach is not acceptable. Don't say, "I am the Boss," and we have never accepted that. You can learn from us too, and it is not just about you having more money.
>
> Roger Bonnet, former director of science, ESA

History of discovery

The Italian astronomer Galileo was the first to look at Saturn through a telescope in 1610. Viewed through his crude instrument, Saturn was a puzzling sight. Unable to make out the rings, Galileo thought he saw two sizable companions close to the planet. Having recently discovered the major moons of Jupiter, he supposed that Saturn could have large moons, too. "[To] my very great amazement, Saturn was seen ... to be not a lone star, but three together, which almost touch each other," he wrote to his patrons, the Medicis.

Galileo was even more astonished when, two years later, he again looked at Saturn through his telescope only to find that the companion bodies had apparently disappeared. "I do not know what to say in a case so surprising, so unlooked for and so novel," he wrote in 1612. The rings were simply "invisible" because he was now viewing them edge-on. Since he was already quite controversial, having seen moons around Jupiter and subscribing to Copernicus' view that the planets revolved around the Sun, he decided to keep quiet to avoid being ridiculed for reporting that the ansea at Saturn had disappeared. Two years later they again reappeared, larger than ever. He concluded that what he saw were some sort of "arms" that grew and disappeared for unknown reasons. He died never knowing that he had been the first to observe Saturn's rings.

Nearly half a century later, the Dutch scientist Christiaan Huygens solved the puzzle that had perplexed Galileo. Thanks to better optics, Huygens revealed in 1659 that the companions or arms decorating Saturn were not appendages, but rather that the planet "is surrounded by a thin, flat ring, which nowhere touches the body." His theory met with some opposition, but was confirmed by the observations of Robert Hooke and the Italian-French astronomer Jean-Dominique Cassini.

While observing Saturn, Huygens also discovered the moon Titan. A few years later, Cassini added several more key Saturn discoveries. Using still better telescopes, he discovered Saturn's four other major moons: Iapetus, Rhea, Tethys, and Dione. In 1675, he established that Saturn's rings are split largely into two parts by a narrow gap, known since as the "Cassini Division." In the nineteenth century J. E. Keeler, pursuing theoretical studies developed by James Clerk Maxwell, showed that the ring system was not a uniform sheet but made up of small particles that orbit Saturn.

And some 330 years after their time, Cassini's and Huygens' namesakes took a much closer look ...

One feature of how the mission was funded greatly assisted cross-cultural cooperation. NASA had centrally appropriated funds, voted for annually by Congress and inevitably fought over by aspiring recipients. In contrast, the *Huygens* probe was financially secure for the entire length of the project because ESA approves its missions for their whole duration, while *Cassini* had its funding reviewed and appropriated annually. Moreover, each European scientist came with a budget attached to her or his participation. American scientists on a team would have to be paid for, but European scientists came "free," the funds to pay for them contributed by their home countries. Proposed scientific experiments were judged on scientific merit but also on cost, so an American principal investigator[13] could put in a low estimate and add European talent if the proposal was accepted. It was a happy conjunction of two different funding systems, enough to overcome any lingering prejudice against European science.

The timing of ESA's 1988 decision to go ahead, the year before Congressional approval of NASA's partnership in 1989, was also important. With the decision to go ahead with the *Cassini*, Europe had taken a bold step forward, even if the mission was contingent on later American funding and even though many American scientists and engineers had been involved in the development and presentation of the mission design. Yet, no one could now argue that Europe was a junior partner. ESA was showing NASA how it wanted to be treated in the future. This had to be a partnership in every sense of the word.

Apollo's dying dream

These inspiring and promising plans had to engage with an American political system seriously "hung over" from the days of John F. Kennedy. Space projects had to be funded lavishly, but for reasons that had very little to do with planetary science per se. Space was a celestial Rorschach test.

> After the *Ulysses* cancellation, we worked very hard to make sure ESA would not again be blindsided. You have to have trust and good relationships. International cooperation is all about personal relationships, and NASA has not always been that strong in knowing how to deal with international partners.
>
> Len Fisk, former NASA associate administrator for space science and applications

Space is a great emptiness into which we often project our dreams. Not unlike the cloud-shaped blots of ink used by psychiatrists to elicit our fantasies that were first designed by Hermann Rorschach, space invites us to imagine vast and limitless opportunities.[14] It has meanings for nearly everyone, but these can be ambiguous and sometimes incompatible. For some people, space is literally heaven, for others unknown worlds, for still others it has a host of metaphorical meanings – freedom, opportunity, limitless vision, infinite possibilities, burning curiosity, and search without end.

Because space missions are so symbolic of multiple goals, many of them not articulated, and because space has both physical and metaphorical meanings, it becomes quite hard to know "who wants what." Hence space agencies are prone to a certain confusion of purpose. The more excited everyone gets about space exploration and what it means, the more lavish are the appropriations. Yet the range of disparate goals that space agencies are expected to fulfill also becomes wider, until it becomes virtually impossible to satisfy every constituency.

Consider the *Apollo* program. Its symbolic value dwarfed its contribution to science. Kennedy had vowed to overcome the "missile gap" and negate the supposed advantage of the Soviets in rocket engines and hence intercontinental ballistic missile (ICBM) delivery. Space exemplified Kennedy's "New Frontier" and evoked memories of Western expansion. These were the Camelot years of questing knights and Arthurian legends, as portrayed by the contemporary musical.

Were Kennedy to win the race in the sky, this would symbolize world leadership. Unaligned nations could watch a peaceful demonstration of whose science was pre-eminent and move towards the Western camp. *Apollo* was actually conceived in the hours and days following the Bay of Pigs disaster in Cuba. It is probably the most ambitious project ever designed to save a President's credibility. Much will be said about paradoxes in this book, and this was one. Kennedy's memo to Johnson demanded a spectacular announcement within 48 hours concerning space. James Webb of NASA was ordered to "dream big enough to get us ahead."

The Soviets had actually reached the moon with *Luna* as early as in 1959, and made a soft landing using *Luna 9*'s robots in 1966, so the robotic (unmanned) race was already lost by the time *Apollo* started. One unfortunate legacy of the "race" was that the United States was obliged to pretend that robotic missions with science instruments were no feat at all; only talkative astronauts making a "giant leap for mankind" really mattered. This introduced a bias into space programs in which human programs had the highest priority and science came second. It became the only way for the United States to "win," and the best way to get money out of Congress. Having said that, without these technological achievements in human space missions, there would of course not be a space science program either.

Americans are strong believers in the separation of powers between the President, Congress, and the judiciary, but this separation has to do with existing, earthbound activities. In outer space lay powers yet to be shared, and if the executive could lay claim to these on behalf of the American people, the institution of the presidency would be greatly strengthened. A few years later Arthur Schlesinger was to call this the Imperial Presidency, the attempt to appeal over the heads of Congress and the judiciary directly to the American people (Launius and McCurdy 1997). Space was the "high ground," not only in world politics, but for domestic presidential ambitions.

The extraordinarily wide appeal of the initial space program had much to do with it being a proxy for enhanced military power. After all, the initial panic over the "missile gap" had everything to do with ICBMs and their ability to take out New York or San Francisco. The US lead in nuclear weapons was of little use without the means of delivering these. For American liberals, space involved the dedication of rockets to peaceful purposes by a civilian agency. For conservatives, space involved rockets of sufficient range to hit any Soviet city in an emergency. For liberals, the stellar spectacular was an alternative to the cold war heating up. For conservatives, it was a way of supplementing the defense budget while developing systems that could always be used for military purposes if necessary. No wonder that both wings of both parties voted overwhelmingly for NASA's appropriations.

In truth, the US government has long subsidized high tech, but it always needed an excuse. In the First World War it supplied the Allies' war effort before finally joining the conflict; in the Second World War it out-produced the German and Japanese war machines; and in the cold war the US government sponsored scores of leading-edge technologies vital to defense and space. It subsidized "troop carriers" which foreshadowed the US airline industry. Space was one more valuable excuse for what might otherwise be seen as socialism. Space was "out of this world" and thus beyond economics and markets.

The civil rights policies of Kennedy and Johnson were not popular in the US South, but spending for space in Texas, Florida, and other southern states won great support. Hence space appropriations first became a way of keeping the southern states Democratic, and then a way to pull them into the Republican camp when first Nixon and then Reagan became President. For Nixon, the astronauts were "the right stuff" – disciplined, cheerful, obedient, clean-cut, and regular – very unlike the demonstrators outside the White House as the Vietnam War wound down to disaster. It was Nixon who got to congratulate the astronauts when they reached the moon in 1969. Yet by that time "space" had lost its liberal supporters; the extravagant hopes raised by the *Apollo* program could not be met, and the political parades had moved on even before the landing was finally achieved. It was now downhill all the way. NASA administrator James Webb resigned when NASA's budget was cut to US$3.9 billion. In the same year, 1968, a Gallup poll found that 53 percent of the public were opposed to a human mission to Mars. The bubble of enthusiasm for space had burst. Human spaceflight had little to show but a circus; by and large it had been a wonderful spectacle, just as barnstorming was in early aviation, but hardly more productive (Roland 1989).

Throughout the 1970s, the US space program gradually lost more and more of its funding, and was given an ever-dwindling share of GNP. Moreover, science was losing out to engineering. The purpose was increasingly to reach chosen targets in the sky rather than to learn about them, an emphasis that increased with Reagan's "star wars" policies.

When Congress was given a choice between funding planetary science and Space Station *Freedom*, it chose the latter.[15] Real science could not compete with symbolism.

In such an atmosphere, a project conceived by scientists would have little respect. But in the aftermath of the 1986 *Challenger* disaster, in which all the astronauts were killed before a large audience, space science was given a new chance. The Space Shuttle was grounded and – seemingly paradoxically – NASA received its first major boost in funding since the mid-1960s, the purpose being to show that neither President nor nation was daunted by catastrophe.

But for the time being, better funding could not immediately be translated into human space flight, and this opened a welcome window of opportunity for large missions such as *Cassini-Huygens*. Moreover, the Soviets embarked in the 1980s on a vigorous robotic science program to Mars and Venus,[16] while the United States launched *Magellan* to Venus and the long-delayed *Galileo* to Jupiter in 1989. For now robotic missions had become the new arena for competition, which was still the implicit motivation, and it started to look as if the Soviets were in the lead again.

When ESA had conditionally decided to go ahead with the *Huygens* probe in 1988, NASA was in fact being put on the spot. It had to make up its mind and persuade Congress. The *Cassini-Huygens* project went against its grain: NASA had always been in charge and now the pressure was coming from below and from Europe. But Lennard Fisk, NASA's associate administrator for space science and applications in that period, was an early backer, as were the key influencers at the Jet Propulsion Laboratory (JPL). One of those was John Casani, a top solar system engineering veteran at JPL who played initially an important role in the creation of *Cassini-Huygens* on the American side of the Atlantic Ocean. If those involved in the robotic space world have any counterpart to the "right stuff" astronauts, it is John: tough as nails and determined to build the most *reliable* yet *advanced* space missions.

This was to be a JPL proposal, since getting to Saturn and orbiting the rings would be its new engineering challenge. Len Fisk began talking to congressmen, John Casani worked with the staff on the Hill behind the scenes, and both made sure the scientists became instruments in the lobbying campaign. Toby had a video made of Daniel and himself, and went to Capitol Hill to lobby, something he had never thought he would ever have to do. On one of his many trips to Washington he happened to be with a professional lobbyist. "When the aide to the senator heard from the lobbyist that the same company that was proposing to build components of the spacecraft had a plant in his state that manufactured Mason jars, he became all ears."

Another complication was that NASA felt half-committed to a comet rendezvous asteroid fly-by (CRAF) mission. Roger Bonnet and his ESA Science Programme Committee had made it clear that if *Cassini-Huygens* was postponed to later years, then ESA's offer to fund *Huygens* and European scientists would be withdrawn. Wes Huntress, who by now had worked himself up to project scientist for NASA, John Beckman and several others at JPL and NASA HQ proposed that CRAF and *Cassini-Huygens* share a single launch platform and propulsion system through the *Mariner Mark II* design.[17] Even though NASA's administrator (James Fletcher at that time) was on record as viewing *Cassini-Huygens* as far more interesting and important than CRAF, his science deputy Len Fisk found the CRAF–*Cassini* combination an appealing way to appease proponents of both missions in one proposal, and led the way towards approval from the White House, the Office of Management and Budget, and Congress.[18]

An improbable ally was also the US Air Force. Its staff had followed developments closely and (unlike their own work, which was classified) they were allowed to discuss this project, which they did to anyone who would listen. The Air Force would welcome funding for an upgraded Titan IV rocket.

In any event, times were favorable for a large science mission like this one. After the *Challenger* disaster, Congress and the White House were favorably disposed towards funding some bold science initiatives that would make a statement to the world, and not least to the Russians. The United States' motivation was to show its technical prowess, and science would also be an excellent demonstration of superpower strength. Major proposals such as the Chandra X-Ray Observatory, the Earth Observing System, and *Cassini-Huygens* were looked upon favorably.[19] Even though the congressional fights about the Space Station at the time appeared to be a last threat that had to be overcome, the CRAF–*Cassini* proposal was in the end accepted by NASA, and Congress approved a US$1.6 billion outlay and voted the first appropriation for it in 1989.[20] The mission was on its way.

The funding crisis

It was on its way indeed, yet not out of the woods. What followed was a demonstration of how quickly the political scene can change. By 1990 the Soviet Union was collapsing, so the necessity for an arms and technological race had suddenly weakened. Japan's commercial challenge to America's hegemony was near its height. The International Space Station *Freedom* was proving very expensive, and a source of international contention due to NASA's unilateralist approach to important decisions affecting its international partners.[21] The economy was heading into recession and a budgetary crisis was around the corner. Whereas European agencies vote funds to cover a whole mission, Congress "reconsiders" every year. What was *Cassini-Huygens*, a huge government-sponsored project, doing for the economy and President Bush? The payoff was in the next century. Quite suddenly, *Cassini-Huygens* found itself in a space budgetary shooting gallery, in which it could itself become a target.

Despite such perils, it is difficult to make deep cuts in space programs without affecting the constituencies of many politicians. Government contracts are vital to many industries in many states, and when the budget squeeze comes, the suppliers to the space community go into defensive mode. Here that meant: how can we hold on to the largest slice of the scientific pie for as long as possible?

The answer lies in arguing that what you are doing is an incremental step to the largest possible number of future goals and opportunities. NASA appropriations are typically presented as stepping-stones to many conceivable futures.

Lynn Ragsdale, professor of political sciences at the University of Arizona, points out that in the Reagan and Bush years, "there was *no national space policy*" (Launius and McCurdy 1997, emphasis in the original). There were no vision, no strategy, and no long-term goals. There were just various ad hoc measures designed to attract continuing funds and stop programs with political constituencies from being canceled. This was what was meant by "incremental advances with details to follow."

Ragsdale distinguishes "primary policy," the stuff of vision and innovation, from "ancillary policy," the stuff of increments, continuation, and policies or tactics. Ancillary policy asks, "What can we afford?" and "How can we sell it?" It does not ask, "What should we do and why?" The manifestation of this ancillary policy was the Shuttle, an incremental step to which everyone could agree because it was a means to goals no one could agree upon. Without a reusable vehicle you could not reach the Skylab or *Freedom*, and you could not make running repairs to whatever else was put in orbit, like the Hubble telescope. The Shuttle was intended to rent itself out to recover its costs – an aim never achieved. It was expected to be cheaper than using an expendable vehicle, a hope that turned out to be false.

Whoever thought up the name "Shuttle" must have been dreaming of a more prosaic existence. The word, as those who coined it intended, conjured up the convenience of short-haul flights. The Shuttle, Skylab, and the International Space Station, and now the plans for Mars again, were all stepping-stones to unspecified ends, suggested because they stood a chance of garnering supporters and were steps towards possible future goals. The advantage of a stepping-stone approach is that you can claim that each step vindicates and continues past policies while bringing new policies nearer, without stating what these policies are, lest they be attacked. You can slow down the introduction of steps so that even cost-cutters like them. What you sacrifice is vision, purpose, excitement, commitment, continuity, perseverance, and of course science, which needs all of these.

> The whole point of leaving home is to go somewhere, not to endlessly circle the block.
>
> Wes Huntress, testimony to the Science Committee of the US House of Representatives, 2003

Ancillary policies are also the result of long-term institutional conflict built into the struggle between NASA, the House, the Senate, and the Presidency. Battles won constitute advances by which one makes haphazard progress. Ambitious long-term goals make easy targets for opponents. Ragsdale refers to "A conglomerate of semi-feudal, loosely allied offices with considerable independence from each other [which] battle over incremental changes."

Third, ancillary policy also acts to deflect criticism of cost overruns. Since everything is a step to something beyond, you will waste previous allocations if you stop now, and will render future steps impossible. Slowing everything to a crawl can control costs. What was to take three years must now take five. The Shuttle's cost per flight escalated from US$10 million to US$94 million. Of the 580 missions scheduled between 1983 and 1994, just 113 were achieved. Estimates seem to vary widely, but the sixteen nations participating in the International Space Station program are expected to have spent US$100 billion over 30 years to develop and operate the station.[22] This is a big price tag for a lot of international animosity and largely for the purpose of learning how to live in space, something the Russians had been doing for years. Projects are given low projected costs in the hope that they will gather Congressional, scientific, and popular support on their way to completion. Reasons are invented along the way, like the prospects of manufacturing in space. In truth, the Shuttle was never economic, and the Russian and European expendable launch vehicles (ELVs) beat it easily on price. ELVs save money and save lives.

In contrast to the ancillary policies, where every project claims to be necessary to every other and just a cautious increment, *Cassini-Huygens* stood by itself, visionary, definitive, very different, and suddenly out of touch with its times.

As of 1989, some time before cuts were demanded and "de-scoping" began, John Casani took charge and made some immediate changes to involve new technology from industry and to reduce the craft's complexity. *Cassini-Huygens* might never have survived but for this JPL veteran of many solar system robotic missions. He is a gruff but highly respected engineer and a brilliant tactician, all of which he would need. He also knew who had contributed to those successful missions. The engineering team he assembled was second to none, as many we spoke with confirmed.

Although Congress had only just voted the appropriation for CRAF–*Cassini*, in less than two years the administration backtracked and canceled CRAF, and ordered major cuts in the *Cassini* spacecraft design as well. The Savings and Loan crisis was continuing, which eventually cost all US tax payers US$1,000 apiece, and President Bush had invited Americans to "read my lips, no new taxes," a pledge that was promptly broken. The Democrats were saying "It's the economy, stupid," and so it was. It soon became clear that canceling CRAF was going to save very little, since it shared the costs of being launched with *Cassini-Huygens*. The savings were less than a quarter of the cost of both missions, and it looked very much as if *Cassini-Huygens* would be next for the chop. The Europeans, in fact, were convinced that the mission was in serious jeopardy. The project's Review Board added its weight by calling the craft as-is an "inoperable, overweight, and underpowered" design.

It was at this point that Dick "Spe" Spehalski – the new Cassini project leader selected by John Casani[23] – and John himself acted decisively. They knew that demands for even more stringent cuts were on their way, and so began the exercise of "restructuring" the entire mission, spurred on by Len Fisk and the NASA leadership. There was still time to save the science part of the mission if they started to plan immediately. To start shedding instruments would cause a storm among the science constituency, soon to swell to 250 experts, yet Spe had to make the very difficult and still criticized decision to remove the rotating and scanning ("scan") platforms, which would have allowed the instruments to be pointed at their varied targets. Many millions could be saved by bolting the instruments to the sides of the craft, although now the entire craft would have to be turned while orbiting Saturn to align the instruments as needed. There was much anger and bitterness on all sides, but things were about to get even tougher. Yet many now say that these three leaders saved the mission from being canceled altogether.

Not only had the craft's design to be "de-scoped" as that was called, but to save more money, Spe, with John cheering him on, had to institute an entirely new way of working at JPL to reduce costs further. He created integrated product teams; co-located staff in an interdisciplinary way; allocated non-negotiable budgets to all engineering teams and the contractors; instituted new ways of tracking project deliverables; and also kept the scientists working on the development of their instruments. There is more about all of this elsewhere in this book; what needs to be known now is that drastic steps had to be taken, and that new, creative ways of "doing business" at JPL saved the mission from the Washington axe. John Casani merely spelled out the criteria to Spe and his team: first, it must work; second, it must meet the new budget guidelines; and third, everything else follows. The scientists knew they had to coalesce, as there would not be a Saturn party at all otherwise. They did, albeit grudgingly.

The skepticism of Dan Goldin

In 1992, the incoming administrator of NASA, Dan Goldin, was already busy reorganizing the agency.[24] On August 22 of that year, the *Mars Observer*, with a heavy and valuable scientific payload, disappeared from radar screens and has never been traced since. While no one could blame Goldin – it had been launched before he took up his post – he felt its loss acutely, and it confirmed his dislike of "big science."

Goldin was a man of the moment, and the moment was all about cutting costs, doing more for the economy, being businesslike, and ensuring regular payoffs from invested funds in the manner of a stock portfolio. His mantra was "faster, better, cheaper." There were to be more missions, cheaper missions, with quicker science returns coming back to the nation. For the cost of *Cassini-Huygens* he could have 20 such missions, spreading the risks and preventing NASA from becoming hostage to huge errors. With this mission things could entirely go wrong, but were they to go right the rewards would accrue to his and his president's successor. On that project he was risking ruin for no tangible reward. He dubbed *Cassini-Huygens* "Battlestar Galactica" after a contemporary science fiction movie, and made no secret of his hostility towards it.

In fact, "big science" was not helping him at that moment. The International Space Station was a cost and design nightmare; the Shuttle had promised economy and frequency but was achieving neither. The Hubble space telescope was in trouble and had to be fixed. *Cassini-Huygens* was the total antithesis of Goldin's new mantra, more expensive, long drawn-out, and decidedly worse. The press began to fill with articles about "over-indulged" planetary scientists.

On Dan Goldin's first visit to Europe, both Roger Bonnet of ESA and Daniel Gautier came away from meetings with him convinced that he and/or Congress were going to cancel the mission. ESA and the European planetary scientists began to communicate with everyone they knew in the United States. Contact was established with the State Department by the ambassadors of many ESA member states, to express dismay at the likely loss of US reputation in Europe if this were done. The director general of ESA wrote directly to Al Gore, who as Vice President had responsibility for space exploration (see the following box, and full text of the letter in Appendix C). David Southwood, who had been a neighbor of Dan in California, made a personal appeal.[25] Carl Sagan, author of the book and television program *Cosmos*, wrote to a large number of high-level contacts, pleading the cause of science. The Italian scientists and ASI leadership were particularly indignant, and reportedly told NASA that Italy's participation in the International Space Station was at stake.

The frantic lobbying appeared to have forestalled cancellation, but then Goldin continued to demand a series of reviews of *Cassini-Huygens*. Such reviews were even forced on the Europeans, who thus would have their competence judged by the Americans.

Whether Goldin wanted the mission aborted by other means or in general, we do not know, although many believed this – certainly in Europe. Every review, however, resulted in high praise, so that in fact the case for canceling such an outstanding mission weakened on each occasion. Eventually *Cassini-Huygens* would make it safely to its launch date, October 15, 1997.

The nest of paradoxes

So the mission survived, but only because top scientists and engineers became politicians: lobbying extensively, pulling every string, fighting for their chance to go the distance, in an environment where horizons were shortening all the time, treating space like quarterly business reports.

Daniel Gautier, Wing Ip, and Toby Owen had proven themselves not simply as interdisciplinary scientists, but also as socio-political organizers of an international constituency that never gave up. The *threat* of being reviewed and even axed by Goldin was turned into an *opportunity*. Instead of letting politics and the timidity of ancillary policy wreck the mission, the scientists and engineers developed an international politics of their own. We agree with Daniel, Wing, and Toby that humanity as a whole is elevated, united, and enlightened by a whole-earth enterprise of this kind.

These were to be the first of many paradoxes this mission confronted, grappled with, and reconciled. In our own quest for a larger meaning, it is to the nature of such paradoxes that we now turn.

Excerpts from a letter of 13 June 1994 from Jean-Marie Luton, Director General of ESA to the Honorable Albert Gore, Jr., Vice President of the United States.

> In making the commitment to participate with the U.S. in 1989, ESA oriented its overall space science programme in order to select this cooperative project, rather than opt for one of a number of purely European alternatives. ...
>
> To date, the Member State Governments of ESA have committed around $300 million to our portion of the mission ... of which two-thirds have already been spent, and have committed to a further expenditure of around $100 million. ... These figures do not include the approximately $100 million contribution of Italy via a NASA/Italian Space Agency bilateral agreement.
>
> Europe therefore views any prospect of a unilateral withdrawal from the cooperation on the part of the United States as totally unacceptable. Such an action would call into question the reliability of the U.S. as a partner in any future major scientific and technological cooperation.

Source: US–European Collaboration in Space Science; National Research Council and European Science Foundation (1998) – Joint Committee on International Space Programs; National Academy Press, Washington, DC.

2
The power of paradox: have values a logic of their own?

> *The test of a first-rate intelligence is the ability to hold two opposed ideas in the mind at the same time and still retain the ability to function.*
> F. Scott Fitzgerald (1896–1940), American novelist

In the last chapter we encountered quite a number of paradoxes. The mission dedicated to Apollo, the god of the Sun, fount of all knowledge, was cobbled together in great haste as a face-saving exercise following the Bay of Pigs fiasco. Amid the Arthurian legend of noble knights in quests for the sublime were some pretty sizable pork barrels. While space exploration is ostensibly dedicated to civilian science, there are some very unscientific, very military and political reasons for supporting it. Do we seek humble discovery, or to strut on a world stage? Is it knowledge we seek, or symbolism? Are astronauts intrepid explorers of genuine independence or Buck Rogers look-alikes portraying comic-book heroism and elevated PR?

We attributed the success of *Cassini-Huygens* to the creation of politics that actually championed science; to scientific ideals known to be realizable; to a diversity of nineteen nations and multiple disciplines, which achieved an astonishing unity of purpose; to the weaving of many sciences into the interdisciplinary synthesis of systems science; to giving Europe and the United States distinct roles in the mission; to paying Europeans in such a way that it cost US-funded teams nothing to include them. We saw that planetary science was "weak" and "threatened" when asking politicians and public for support, but how, with international help and political savvy, it grew strong and seized the opportunity.

All these values and more are *paradoxical*. The values in the first paragraph were not reconciled to the satisfaction of those concerned, and so impeded space exploration. The values in the second paragraph *were* reconciled, and this contributed to the success of *Cassini-Huygens* and its exceptional feats.

Are these paradoxes coincidental? Paradox has often been used for literary flourish. Some famous book titles are paradoxes: *Point Counterpoint, The Chrysanthemum and the Sword, Darkness at Noon, Silent Spring,* and so on. Are they simply transitory on the road to reason, shadows that vanish before enlightenment? Or can we find within paradox a coherent logic? What we are attempting with this book is a logic of paradox in and of itself, a way of making sense out of highly innovative learning and growth dynamics. This logic applies not to the physical universe, but to the realm of culture, values, and our social processes. Through this new logic we seek to explain how the protagonists of *Cassini-Huygens* rose to so difficult and demanding a challenge. In this secular age we seem to have lost our bearings over what might be truthfully regarded as magnificent. Even our gods appear to take sides. Can nothing be rescued from the relativism that sees all realities as partial, all cultures as legitimately different?

More than 20 years ago, in 1982, Peters and Waterman wrote their bestseller *In Search of Excellence*. A major theme was that excellent organizations "manage values." We thought of calling our book *Excellence Discovered*, but perhaps that would have been too inward-looking and too derivative. For what we observe is that powers that lie within people need to be called forth by something of limitless aspiration "out there." Like our heritage of religious art, the eternal evokes transient humans to surpass themselves. Dedicating your selfhood to a larger goal sustains that self and renders it transcendent.

You discover just how much is possible when you attempt the impossible. The price of unalloyed success is to look failure in the face repeatedly. The path to excellence is strewn with shortcomings, thousands of them.

What distinguishes the capacity to learn rapidly, culminating in superlative performance, is the capacity to confront and to resolve *value paradoxes*. In an earlier paragraph these included inner-excellence–outer space, possible–impossible, success–failure, and excellence–shortcomings, but these are only a foretaste of far subtler dilemmas and paradoxes, which will be examined in the later chapters of this book.

A word should be said about paradox itself, and how we have been trained to fear and distrust it. The first major science used to examine our universe was astronomy, followed closely by physics. These taught us to regard heavenly bodies as objects occupying public space, out there in a vast universe, isolated from each other and totally detached from us. None of our stories, myths, or beliefs about the stars and planets turned out to be true. They do not influence our lives, as astrologers still pretend, and they are indifferent as to when we were born. There is nothing god-like about any constellations, and they have no known inhabitants.

Because astronomy, physics, and the later "hard" sciences turned out to be so successful, those who thought about human values tried hard to be "scientific" as well. They assumed that values were also isolated objects, akin to stars and planets, varying in size from large to small. Hence, the larger the virtues you possessed are, the more unity, competitiveness, certainty, loyalty, realism, logic, cooperation, dissent, doubt, idealism, and creativity you exhibited, the more likely would your mission be to succeed.

But there is a problem with this list of virtues. Might not greater *unity* be strained by *diversity*? Might fiercer *competitiveness* make you less *cooperative*? Could your *loyalty* be in question if you *dissent*? Could *realism* blunt your *idealism*? What if the *logic* you are using is rendered obsolete by *creativity*?

Within this list are a series of contrasts. To speak of "added value," as economists and business managers do, is to assume that values *add* to each other. In fact, they could subtract from each other, just as doubt subtracts from certainty and dissent from loyalty. Scandal in business has shown how confidence in well-performing companies can be undermined by the doubts and waning loyalty of customers.

Aristotle taught us that all things must be classified into "A" or "not A," and that you must not contradict yourself. To say that "competition is cooperation" or "doubt is certainty" seems contradictory and hence nonsensical. How can you be both diverse *and* unified?

The fear of contradicting ourselves has led many intelligent people to avoid all moral judgements in learned discourse. We are especially frightened of appearing irrational and of confronting dilemmas or paradoxes, from which we will be forced to make an undignified exit. To get an answer that contradicts your proposition advertises your mistake or ignorance. If you expect to find mostly methane on Titan and find mostly nitrogen, your previous assumption must yield to the evidence. Paradox is overcome and a singular reality is restored. We do not like anomalous findings.

To posit the existence of paradoxes also makes it difficult to verify scientific propositions. If your hypothesis is confirmed by what you expected to find, but also confirmed by its negation, then is your theory falsifiable? If you were wrong, how would you know? We have entered a world of smoke and mirrors, wherein propositions and negations are both true.

No wonder that large sections of the scientific community have given up on moral discourse altogether. If they have moral convictions, these are said to be private, apart from their professions, matters of personal belief. Value judgments are held to be without "testable meaning" beyond subjective preference. To say something is "good" is no more significant than saying you like strawberry ice-cream. You are making "exclamations of preference" about your moral taste buds.

We do not dispute the role of objectivity in science and engineering, or common rules of classification. In an objective universe of things observable in public space, focusing our lenses on discrete units is clearly justified. What is *not* justified is treating values as if they were things or objects. Indeed, it may be the delusion of our age. It is easy to see how common sense made this error. Objects have names, such as table, lamp, chair, sofa, and moon. Values also have names, such as certainty, doubt, unity, diversity, dissent, and loyalty. Hence each name is thought to correspond with a value, or essence of that value. There must be a thing called certainty, just as there is a thing called moon. If only we could locate these elusive "essences." But we never have, and we never will. Values are not things at all, but are *differences*, so:

Certainty	*is in tension with*	Doubt
Unity	*is in tension with*	Diversity
Loyalty	*is in tension with*	Dissent

What happens is that the people, teams, and organizations move *between these differences*, for example from systematically doubting to becoming more certain, using international diversity to create a more powerful unity, demonstrating fierce loyalty to the mission by dissenting on issues of perceived peril. On such occasions the six values at the end of these three continua are synergistic (from the Greek *syn-ergo*, "to work together"). In many, perhaps most situations, doubt and certainty do negate each other, as do unity and diversity. Consider how a scientist succeeds. She or he *doubts* propositions in order to become more *certain* of these, considers a *diversity* of possibilities and approaches so as to forge a better *unity*, and *dissents* on occasion from fellow scientists out of a genuine *loyalty* to them and the joint mission. In this way a perceived *weakness* can become a *strength*, and a potential *threat* an *opportunity* to improve.

Note that regarding values as separate "things" will get us into trouble. Pushing the idea of risk to its logical extremity can lead to despair. Magnifying diversity will make it unmanageable, a veritable Tower of Babel. In such cases value is not created but destroyed. Great science or great enterprise needs doubts that are tested to make us more certain, diversities explored to create new unexpected unities, and dissent proved by subsequent events or findings to have been motivated by loyalty.

So how do contrasting values achieve synergy? They need to be managed and hence reconciled. Failure to reconcile them is probably the most common outcome. Those who doubt most are typically racked by uncertainty. Those who seek more and more diversity may find themselves inundated by information they cannot assimilate. Such situations testify to the fact that values are often in serious *imbalance* – too much dissent, too little loyalty, too much doubting, and insufficient confidence. But something very interesting happens when values are well balanced and evenly matched. All of a sudden these learn from each other and become *mutually enhancing*, so that the doubts you have entertained make you more certain, while the sheer number of diverse inputs helps produce the unifying breakthrough. The contrasting values are now being reconciled.

A writer who has put his finger on this dynamic is an American psychologist with the near-unpronounceable name of Mihaly Csikszentmihalyi.[1] He has spent a lifetime studying a phenomenon he calls *flow*, an experience somewhere between great happiness and intense absorption in shared or individual tasks. Suppose that a team of people is engaged on a joint task that constitutes a challenge. To this challenge they bring a combination of their skills. In the vast majority of cases challenges and skills are mutually subtractive. If challenges are greater than skills you fail, amid a whirlpool of anxiety. If skills are greater than challenges you may also fail, because the team is insufficiently stretched and may suffer from boredom. Boredom and anxiety are common experiences in many organizations. This is illustrated in Figure 2.1.

Yet something very different occurs in those cases where challenge and skills are finely matched. Let us suppose the challenge is just inches beyond the level that skills have attained. Might we not expect that people with these skills would strain every nerve and sinew to come up to the mark? Might not the challenges excite them to higher levels of energy and performance? They might rise to the occasion, fascinated by the prospect of mastering a new problem and intensely absorbed by this goal. Might the challenges educate them to lift their skills to levels higher than ever before?

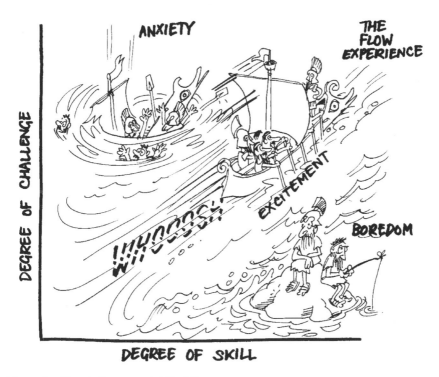

Figure 2.1 The challenge and skill difference

It could work the other way around. If the challenge has been matched by available skills, why not raise that challenge? Cut costs even more, widen tolerances for the extremes of temperature the spacecraft must face? Just as a high jumper raises the bar after successfully clearing it, why should not a team agree to beat its own target? The team is interested in discovering just how good it can be. A spacecraft this complex cannot be too fail-safe as far as engineers are concerned, or its mission too curious as far as scientists are concerned. Engineers need to get you there; scientists always want to know more and are never content.

In experiences of flow, the borderline between skills and challenges tends to dissolve. You solve problems, beat records, and surpass your previous performances for the sheer exhilaration of that experience. Champion swimmers will tell you that they are one with the water, as the skier is with the mountainside, and the runner with the race. You break through a pain threshold, and suddenly, what earlier slowed you down now spurs you on and propels you in your desired direction.

In the case of this book the challenge was nothing less than journeying to Saturn, orbiting through its rings, and landing a probe on Titan. With such an exciting and ambitious challenge, we might expect the accumulated skills of team members to go into overdrive. This was the mission of a lifetime, a legacy to leave to an exploring world full of curiosity about this fascinating planet, studied for over 2,000 years.

While we agree with Csikszentmihalyi that challenges and skills can form a virtuous circle of mutual excitation, we do not believe it ends there. We believe that superlative performance is driven by reconciliations and flow experiences *among many different paradoxes,* between diversity and unity, doubt and certainty, dissent and loyalty, error and correction, and so on. *Brilliant performances resolve these paradoxes.* Business as usual trades them off or ducks them and so lacks luster, or even leads to calamities. Could that have been the case with the *Challenger* and *Columbia* disasters?

Csikszentmihalyi is not the only expert to point to paradoxes. Business studies have long generated an assortment of these. Paul Lawrence and Jay Lorsch (1986) demonstrated that while the extent to which a corporation's sub-systems are *integrated* and the extent to which they are *differentiated* are often in conflict, creative corporations in complex environments are both more differentiated *and* better integrated.

Peters and Waterman (1982) offered several injunctions, three of which were paradoxical. They advocated "management by wandering about," or indirectly discovered direction; "simultaneously tight and loose structure," and both hard and soft systems. Richard Pascale (1991) wrote of *Managing on the Edge* (of chaos), and that we must admit the chaos of the market into our mindset and then struggle to find order.

Henry Mintzberg (1994) contrasted designed with emergent strategies, finally conceding that strategy is crafted *out of* what emerges. Collins and Porras (1994) used a ying–yang design to suggest the harmony of several organizational paradoxes. They cited Johnson and Johnson's 1941 credo "shareholders come last," and asked us why in this case Johnson and Johnson had returned so much profit to its shareholders over the last 60 years. The reason was simple. You first have to motivate your employees to delight customers, and only then will the funds become available to shareholders. "Last" means "last in the sequence," not "least important." Customers' delight and shareholders' profits *grow together*.

Finally, W. Edwards Deming (1982) has argued that continuous improvement in manufacturing as well as higher quality and lower cost depend crucially on setting up an *error-correcting* system in which errors are pounced upon and eliminated.

All these are potential paradoxes and reconciliations. What distinguishes this book is the championing of *positive paradoxes*. We believe there are a large number of these paradoxes, comparable to a dictionary of synonyms and antonyms. Where reconciliations can be discovered for these contrasting values, they constitute a major force in the social, economic, or technological realm. We plan to show that several of these value paradoxes and their reconciliations propelled the *Cassini-Huygens* mission.

One of the riddles we need to solve is why the *Cassini-Huygens* mission has not shared the fate of many other missions. No one looking at NASA's record over the last 25 years could be uniformly impressed. Three times spacecraft with human beings on board have failed and their crew have been killed, twice with many of us watching. The Hubble space telescope was launched with a mirror that had a spherical aberration so that it initially saw almost nothing. It needed an audacious human space mission to carry out running repairs, at ruinous expense. With the delays in flying the Shuttle as a result of its grounding, the Hubble telescope may have to be abandoned, since further repairs may not be feasible. As for the International Space Station, who knows what its eventual payoff will be?

On robotic missions antennae have failed to open (*Galileo* to Jupiter). *Mariner I* (Venus) and *Mariner 3* (Mars) failed at launch. Contact with *Mars Observer* was lost on its arrival at Mars, just when it was to ignite its thrusters to enter orbit. Its fate is unknown. There was also a mishap with the Mars polar lander in 1999.

The European Space Agency has had trouble too. Recently, the UK-led *Beagle* lander on the *Mars Express* seems to have crash-landed on Mars and we have not heard a peep from it. Some Ariane rocket versions have been unreliable, and several missions have been aborted. Contact with the SOlar Heliospheric Observatory (SOHO) spacecraft – a joint ESA/NASA mission – was lost in 1998 as a result of a number of operations errors. Brazilian space missions have failed three times before launch, on the last occasion killing 40 people.

So what does *Cassini-Huygens* have that other missions lacked? We think its success happened because it confronted paradox after paradox, perhaps unwittingly, and reconciled many if not all of them.

We made a start on exploring this issue in Chapter 1, where we saw that space science had emerged from a long period of being dominated by the "Imperial Presidency" of the United States, foreign policy objectives, Europe's policy of independence, pork-barrel contracting, and cold war technological objectives. In short, it was driven by everything but its true purpose, the advancement of science in space. At last in the early 1980s, as the cold war eased and space exploration reverted to its original purpose, there was born a genuinely international space-based mission, relatively free of hidden agendas and partisan distortions. Science and the curiosity of scientists had a chance they had never had before, on a scale previously undreamt of.

On the subject of dreams, we saw in Chapter 1 that rarely – if ever – was an idea and the *idealism* it evoked so fervently embraced by so many scientists and engineers in so many nations, all determined to make it *real*. Rarely has *politics* so brilliantly enhanced *science*, instead of subjugating it. Once the mission was under way, the sheer enthusiasm and dedication of this lobby could not be denied. Its very name evoked the scientific feats of France, Italy, and Holland in the seventeenth century, when the European enlightenment was born.

Idealism is not enough, though, and it has on occasion proved catastrophic, yet not in this case. On June 30, 2004 the ideal began to be realized, as the *Cassini* orbiter was inserted into its rotational paths around Saturn, and pictures and movies are now being beamed back to Earth, whose clarity and number are without precedent. We can afford to aspire to great ideals, provided we are very practical and realistic about our political means of achieving these. Like the contrast of skills and challenges we discussed earlier, ideals and the means for their realization need to be matched and synergistic. An ideal *really* comes to life when those who hold it realize that it is attainable. Science can inform political judgment.

Each of the remaining chapters of our book identifies pairs of contrasting values that presented formidable challenges to the success of the mission. Yet its members found the skills to meet those challenges, so that opposites fused and flowed.

In **Chapter 3** we illumine the paradoxes of *design* and *review*, and of *error* and *correction*. Errors are inexpensive provided they are made in play or simulations, but in the actual mission they can lead to very serious outcomes. Those who make mistakes enjoy a vividness of experience and a sense of drama which are largely missing from flawless operations. Errors provide heart-stopping moments that engrave themselves on memories. This is the most eventful and fastest way to learn. These values can "flow" as one, with reviewed designs and playful errors informing the serious business of correcting. The mission achieves not so much perfection but rather a process of perfecting, continuous improvement via successive approximations.

In **Chapter 4** we look at the paradox of *competing* as individuals, yet *cooperating* as a community. We also examine how an amazing diversity of cultures and disciplines can end up achieving a unity of purpose. The competition among individuals to get their instruments aboard the orbiter and probe on this mission was fierce indeed, as was the struggle for observation slots. But once formally admitted to the team, the mission took off, and the individuals became a tight community, sharing the same fate and destiny. What was evoked by *competing* became a touchstone around which all could *cooperate* in a process that has been called "co-opetition."

Participants were helped in this respect by the fact that Saturn is an interactive, dynamic system. Each investigator with his or her own diverse viewpoint discovered aspects of a system that was unified; the part each played converged on a whole. Those who maintain their faith in the unity of nature may have to struggle on alone for many years, but they know they will eventually encounter and greet fellow pilgrims at a clearing in the woods.

In **Chapter 5** we look at the paradox and the interaction of *engineering* with *science*. This pits the pragmatic against the speculative, builders against explorers, the predictable against the unpredictable, the linear against the circular, and the means of getting to Saturn against the ends of exploration and discovery. Obviously engineering must be done first, or the science cannot begin. But will all resources be expended on getting there, leaving only crumbs for science? Can the pride that comes from hitting your target before a world audience give way to the humility of asking questions of nature, bowing to her verdict, and admitting how little we know, how much we have to learn?

We shall see how a mission initially conceived by planetary scientists had scientific objectives designed into its engineering. While engineering concerns briefly dominated in the lead-up to the launch, as the mission neared Saturn the two disciplines became enraptured by their common quest. Their commitments fused into a shared wonderment.

In **Chapter 6** we look at the public outcry that greeted the use of nuclear fuels aboard the spacecraft. To meet this, the team requesting presidential launch approval imagined several nightmare scenarios, and took elaborate precautions to neutralize each eventuality. But there were also real *crises* within which the mission found *opportunities* to improve, demonstrating an extraordinary resilience and the capacity to respond to disorder with renewal at still higher levels of order. On these occasions it is typically too late to undo mistakes. Rather, a "no blame" culture must devise a creative resolution, and insert this at a later stage in the operations. The recovery from one crisis actually provided more and better science for the mission than did the initial plan.

In **Chapter 7** we describe how *elites* among many planetary sciences and types of engineer learned to treat each other as *equals*, for as long as it took to search amongst themselves for possible answers. Who should wield authority when each is "best" but in a different domain? Who then should lead, and who should follow? This question is unanswerable without the mediation of an inquiring culture. In practice diverse experts "rule by turn," but leave hard traces of their influence and their high standards on mission culture, which is shaped by successive leaders, each leaving a record of excellence.

In **Chapter 8** we take up the only paradox that was not successfully reconciled and could have aborted the mission, that between *simplicity* and *complexity*. NASA's administrator, Dan Goldin, believed in the simplicity of the chain of command. He would give orders, demand compliance, and hold his subordinates personally responsible for any failure. But there was no way his direct reports could pass on marching orders to nineteen nations, two other space agencies including the US Air Force, several Washington agencies, and scores of different institutions. Nor could Saturn be reached by mandating a succession of simple steps. Goldin never forgave this mission for its complexity, but railed against it. Yes, complexity needs to be simplified – but not assailed. Yet curiously Goldin, the arch-skeptic, demanded such frequent reviews that reviewers and reviewed became progressively more certain the mission would work.

In **Chapter 9** we draw out the lessons of this book, not just for space exploration but for outstanding performance in many fields. Values are rather like the primal soup being investigated on Titan. The ingredients and the components of life may all be there, but these have not yet germinated and come alive. Values are not just elements but processes of self-organization, in which one value flows into another to enliven both. Ingredients mixed together in a cocktail shaker may only dilute each other so that errors, for example, subtract from correct answers in a battle of right versus wrong. In contrast, an error-correcting system employs errors to inform improvements and spur corrections. The values form recursive loops that promote learning.

The key lessons include the following. The more diverse are the participants in any enterprise, whether the diversity is of culture, of nation, or of discipline, the more these need to be united by a superordinate goal which transcends that diversity, a vision and purpose so ambitious and magnificent that petty misunderstandings are dwarfed. The more complex a task or challenge, the harder it becomes for this to be encompassed in the mind of a single leader. What is required is a culture informed by the inputs of many experts exercising passing influences. This culture guides, motivates, assesses, and celebrates that combination of excellences which meets the challenges faced by the group. It is not simply what you know but what you do not know. A group capable of learning needs to be steered towards areas of ignorance, to set itself up for the greatest possible surprises.

It is abundantly clear that this mission lacked the usual "motivators" for its participants, such as bonus payments, profits, fame, promotion, high status, beating competitors, boosting the reputation of their nation, and pleasing those in authority. Yet something far more profound drew them on: a search for meaning, an intense pleasure in creating knowledge and order out of ignorance and chaos. Learning and discovery are their own rewards.

Finally in the **Epilogue** we narrate the touchdown of the *Huygens* probe on Titan amid not so much of a flow, but a flood, of experiences, tears, excitements, congratulations, relief, joy, even sadness for those who had not lived to see the day. This we regard as a vindication of Csikszentmihalyi's model. When paradoxes are resolved there is an intellectual and emotional "high" for those involved in the solution, which constitutes a solution-type flow of high energy, a quasi-spiritual experience, as when one value breathes (*spiritus*) life into another.

This book is written to help not only leaders and managers on missions of this kind, but all who must wrestle with the knowledge revolution and the increasing complexity of business in general, and international business in particular. Up to now we have opted for a range of single "motivators." Perhaps the chief of these is money. We find out who is doing really well, shovel money in their direction, and hope they will stay. Whatever the value of this motivation, it cannot explain the superlative performance of *Cassini-Huygens*, most of whose protagonists received nothing more than a routine middle-class stipend, higher than they would have got in academia, but not by much.

What about status and public popularity? Some astronauts achieve this, but it tends to crowd out serious science and discovery. The protagonists of robotic missions like *Galileo, Ulysses*, the Mars rovers *Spirit* and *Opportunity*, and *Cassini-Huygens*, are not household names, most of them never will be, and yet they do not complain. Whatever spurs them to activity, it is not the prospect of being famous. So esoteric is the work of most of them that you would have difficulty following them if they tried to explain their contributions. This is not the stuff to make people popular at parties. Perhaps people they meet envy their affiliation with such a mission, but that is all.

Some people hold that "participation" includes untold pleasures, and that people work for the love and conviviality of colleagues. There may be something to this, but problem-solving teams are of brief duration, breaking up after the problem is solved. To make them your "family" is to invite early bereavement; to become attached emotionally is about as wise as an office affair. Participants' first commitment has to be to the mission, not its individual members.

Nor can this mission be seen as a triumph of free enterprise and the profit motive, although several contractors served it well in Europe and America. It was proposed by academics and sponsored by government agencies. It never aimed to make a profit, and may eventually cost the taxpayers close to US$4 billion.

What then is going on? Why do the people we met weep in public, pause to compose themselves, and speak of the legacies they leave behind? Why did they experience episodes of intense pleasure, bordering on joy, and an unforgettable depth of engagement with others, and why did so many of them call this the "highlight of my professional career"? It is our contention that reconciling value differences – so that these flow together in great surges of

> What I think is marvelous about this mission is that it has the seriousness and the ambition of the *Apollo* program yet does far more for science. Surely the best scientists and engineers have a purpose in life, and that is to leave a legacy to future generations. I will see, before my retirement, the result of this mission, but the real value will be bequeathed to my students, my scientific sons and daughters. You cannot be a great scientist without wanting to be a great teacher, and the greatest teachers are those who have been in the thick of things. And what we have to pass on is not just science but hope, an optimism that will sustain our future lives. My upcoming students may be smarter and quicker than me, which is as it should be, but I have had the experience.
>
> *Marcello Fulchignoni, principal investigator,* Huygens *atmospheric structure instrument (HASI), University of Paris 7/Observatoire de Paris–Meudon*

excitement – is inherently pleasurable and self-fulfilling. It may even transcend the self, as it joins into the unfolding growth of human knowledge of the solar system. Some to whom we spoke were even able to articulate that feeling of having touched that great stream of knowledge which is the final destiny of the human mind.

The word "solution" has long intrigued us. At one level it means a liquid; at another level, an answer. Could it be that reconciliations *take liquid forms* in which the paradox dissolves into a new solution? Yet only if we confront these cross-pressures can we find the flow of life itself, the powerful stream partly on the surface and partly below consciousness, but forever pressing on.

Each chapter that follows takes the story further and presents fresh contrasts that give way to confluence. What on Earth is Saturn teaching us? We start with *playing seriously*.

3
Errors and corrections: how to play seriously

> *Research is what I'm doing when I don't know what I'm doing.*
> Wernher von Braun (1912–1977)

Errors and corrections are the first of the major dilemmas we discuss. This issue confronts not just *Cassini-Huygens*, but all enterprises needing to generate new knowledge for new purposes. It also confronts those for whom serious mistakes are unaffordable and potentially catastrophic, and who need the utmost precision and exactitude to fulfill their common goals – agency and political leaders, technicians, hospital staff, pharmaceutical researchers, top executives evaluating large capital investments, and so on.

Space exploration is simply a very demanding example of these circumstances. It requires us to be perfect, or as near perfect as is possible. A single error or oversight may prove fatal. In recent years we have been cruelly reminded of the fact. The O-ring seals aboard the *Challenger* proved too brittle when the temperature dropped. The ascending rocket blew up seconds after lift-off in the presence of a festive crowd. More recently some insulating foam came loose during ignition on the *Columbia*, striking and damaging the heat-resistant ceramic tiles on the leading edge of the shuttle's wing. The mission appeared to proceed as planned but when the shuttle re-entered the earth's atmosphere, the intense heat of re-entry permeated the damaged tiles and destroyed the shuttle from within, so that it broke up over Texas. In both instances the crew perished.

These were "small" defects, which may have been known to mission managers yet overlooked. Were objections by engineers at lower levels overruled, plans to study the matter delayed? Other disasters to robotic missions were even embarrassing. In one case American measurement units were confused with the European metric system. We shall see in Chapter 6 that some of the serious errors on *Cassini-Huygens*, fortunately caught in time, were quite elementary.

So can we be perfect? Can we exercise so much care, precision, thoroughness, vigilance, oversight, and attention to detail, that mistakes do not happen? And if we cannot be perfect, then should we avoid missions of great ambition and complexity? Should we spread the risk among many smaller probes, as NASA ex-leader Dan Goldin used to insist?

It might lower the risk threshold if we never used anything except tried and tested technologies. But this would mean clamping down on creativity and proclaiming a 20-year moratorium on improvements. Many new technologies are more reliable than those they replace. They have been created because the older systems were unsatisfactory. We shall not reach Saturn with obsolete junk.

It is certainly true that people make many mistakes when they are being creative, but they also make mistakes when endless repetition of practiced routines bores them stiff. Routinizing the Shuttle proved disastrous. "Deviance was normalized" (Columbia Accident Investigation Board 2003). The dismal reputations of most bureaucracies suggest that such cultures are extremely error prone – anything to break the monotony! Perfect routines are sleep-inducing.

Along with the fallacy of automated persons who never err, there is the equally fallacious concept that an indomitable force of character will see you through. In fact, NASA had a reputation as a "perfect place," a sort of Camelot, in the days when massive funds were spent on "beating the Soviets."

Such ideals of perfection are fatally flawed. Recycling Leni Riefenstahl's *Triumph of the Will* gets us nowhere, attractive though some find it. The notion of a will to succeed that takes no prisoners and perseveres to crush all opposition will only succeed in destroying that with which it engages. *Cassini-Huygens* did not make it through the unrelenting optimism of can-do confidence, prayer breakfasts, quantum thinking, and "never take no for an answer." Such illusions belong to Captain Marvel and his ilk.

There *is* no immaculate value, no "single-principle imperialism" that can explain the success of *Cassini-Huygens*. Cultures that boast of their unconquerable souls leave a trail of devastation from Vietnam to Chechnya, from Enron to MCI WorldCom. They refuse to be discouraged when things blow up in their faces, and they compound their errors.

The answer to the success of this mission goes to the heart of paradox. You *design* yet you *review* this design, not once but repeatedly. You make many *errors*, so that you can *correct* these; you *simulate* to prepare you for *reality*, and you *play seriously*. Because you have so often *doubted*, you grow progressively more *certain*. You show *friendship* by being super-*critical*, and *support* those who make *mistakes*: that is, the person not the mistakes.

Through the progressive reconciliation of all these paradoxes or dilemmas, you fine-tune a process that is self-perfecting and over time comes closer and closer to the ideal of exploring the Saturn system and accomplishing a mission of awesome complexity. Designs improve by being reviewed. Errors are memorably etched upon the mind before being corrected. Simulations teach you cheaply about reality. Testing your doubts makes you progressively certain, and you prove your friendship by being critical of the shortcomings of colleagues.

Among the beneficiaries of error-correcting systems is, for example, Toyota, recast by Taichi Ohno and W. Edwards Deming in the 1960s.[1] Today the company is worth more than the entire United States automobile industry combined. Peter Senge has called this the Fifth Discipline.[2]

This process of continuous improvement has the following aspects.

1 Design and review.
2 An error-correcting system.
3 The simulating of reality and playing to get serious.
4 The self-repairing mission and the Pygmalion effect
5 Wide tolerances and built-in redundancies.
6 Openness and comfort to be critical.
7 Scrutiny and trust.

As we explore how the project learned in an accelerated way through error and self-correction, we will let the engineering members of the team tell their stories of important aspects of the *Cassini* spacecraft's and the *Huygens* probe's development phase.[3]

Design and review

"This is a business of continual review," Kai Clausen, system engineering and operations manager on the *Huygens* development team at ESTEC, told us. "We are in the business of cross-checks and double-checks. Everyone looks out for everyone else to make sure there are no more mistakes."

"Arguably, this was *the* most reviewed robotic program in NASA history," claimed Dick "Spe" Spehalski. "We had review upon review upon review. You need really rigorous testing to reveal mistakes and failures." Spe seems like a man who does not tolerate mistakes. He makes you think of General Schwartzkopf – congenial and a warm smile, yet he has that demeanor that makes you want to salute him, and a pair of eyes that can stare any soldier down.

Here we should also introduce Hamid Hassan, the leader of the *Huygens* development team.[4] British-born of Pakistani origin, and an ESTEC veteran, Hamid had an enormous grasp of all the details that went into the *Huygens* design and development. He attended most monthly progress meetings and the numerous reviews and testing events with Aerospatiale and its 40 subcontractors across Europe for over seven years, and negotiated hard with them to keep them accountable for on-time delivery and proven performance. Weekly calls with the NASA team at JPL were standard, followed by the team's weekly whisky parties and betting on silly things to win the next bottle.

Since *Cassini-Huygens* looks like it will be the most ambitious and scientifically fruitful mission in the history of space exploration to date, these constant reviews and the mission's overall success may be connected. When we come to examine the specific crises that threatened the success of this mission, we shall see that some problems still escaped the review process. It points at a key aspect of these reviews: the designer of a component still carries the same responsibility for error-free design, and a review does not abrogate that responsibility.

> Hamid brought a bottle of whisky every week. Not that we drank too much, but at least one thing I got from this program was a taste for pure malt whisky. But even after seven years of one bottle a week, we could not cover the full map of Scotland! We had our tensions, but never to a level that it was an issue. We played and worked hard at the same time.
>
> Michel Verdant, Huygens *manager for payload, assembly, integration, and verification*

The process of review started early with the cancellation of the CRAF mission by US Congress. The legislators doubtless hoped to cut the budget in half, now that there was one mission and not two. But in fact the two missions were to have a similar launch vehicle and platform, so canceling one increased the costs of the other. If *Cassini-Huygens* were not to share the fate of CRAF in the next budget round, the project would have to be redesigned, or "de-scoped" to accomplish less. This led to considerable review and redesign, in which costs were cut drastically but the scope to conduct science was cut far less.

The CRAF cancellation preceded the arrival of Daniel Goldin in April 1992, and Goldin made it clear in press interviews, speeches, and asides that *Cassini-Huygens* was an extravagant holdover from NASA's past excesses. He had many people convinced that he intended to cancel the project. Whatever the truth of this allegation, *Cassini-Huygens* was redesigning itself to save itself. The first thing Goldin did when he came on board was order a "Blue Team versus Red Team" exercise for all NASA projects to scrutinize their viability; in addition, he instituted annual and mandatory independent reviews. Later he would demand re-certification of all missions, and *Cassini-Huygens* was no exception.

Constant review was built into the way the spacecraft and its instruments were assembled and integrated. Every component or sub-system had to be delivered on schedule to a larger sub-system, leading up to final integration. Each component and sub-system not only had to be first tested by the unit that delivered it, but also if important enough it would be subjected to independent reviews by non-insiders. It then had to be tested again by the unit that received it and integrated into the wider system. It had to work in isolation *and* in context, and was not officially "received" until the second test had been successful. Hence a single circuit might be tested more than a score of times as a crucial element in an ever-expanding system. At JPL and in Europe, internally and with the main contractors and subcontractors, there were multiple-level quarterly reviews, preliminary design reviews, system design reviews, electrical and mechanical hardware design reviews, critical design reviews, and independent external reviews.

In 1995, Goldin demanded a review by an American team of external experts of the work being done on both the *Cassini* orbiter and instruments under development at JPL, as well as of the *Huygens* probe done by the Europeans. To the latter this was potentially inflammatory, particularly in the context of new American laws decreeing that Europeans were not to have access to any American know-how crucial to defense.[5] American access to European work was to be unilateral. With considerable maturity ESA accepted the demand, while the project team members on both sides regretted that this had to happen. Yet scrutiny by an independent authority could add that extra margin of safety.

Much of the credit for this must to go the reviewers. The team, led by Professor Herb Kottler of MIT's Lincoln Laboratory, reviewed the *Cassini* orbiter and instruments, the work of the Italians in regard to the antennae, and the development and design of *Huygens*. It proved so helpful to all concerned, and the reviewers were so admiring of the work that had been accomplished, that the mission became virtually unassailable. There could be no better example of growing certainty and confidence generated by systematic doubt. Kottler's team said of the European probe review, "Extensive system testing, engineering, and testing and review processes have resulted in a sound design of the probe. The review team identified no major issues, but for a few remaining items."

> The Kottler review in the end turned into a positive experience and created confidence on both sides. We in Europe adopted the external review process after that experience when we had problems with the *Ariane 5* and with the *Hipparcos* missions. Kottler proved to be very competent and his report was both very good and friendly.
>
> *Roger Bonnet, former ESA science programme director*

Not everyone we spoke to praised these constant reviews. They were expensive, time-consuming, and ate into scarce project funds. Some thought that genuine issues might have surfaced sooner, and later crises could have been avoided, if the mission had sorted out its own problems. Others felt that some reviews were so obviously political that faith in the process was harmed. As we shall see, some faults got through nearly *all* of the review processes to create crises. Nonetheless, now that we know *Huygens* has successfully visited Titan, it appears that the most reviewed robotic mission ever may well have a claim on being the most successful for some time to come.

An error-correcting system

One way of thinking of design–review–redesign is as an *error-correcting system*. But why have any bus with errors at all? Are these not our mortal enemies, threatening to ruin the lifeworks of hundreds of scientists and engineers on two continents? Surely the fewer errors we make the better, and the sooner these are stamped out the better?

On the contrary, errors are one of the most important ways we *learn*, and everyone involved with the mission was engaged in a learning race to get everything working at maximum effectiveness by a set of crucial deadlines. If you do not learn fast enough, a fatal flaw may remain hidden until too late. It is a race towards perfection, a goal you never reach but a vital approximation nonetheless.

Errors are *memorable*. They bring you up short and etch themselves powerfully into your mind. To plot a better path to your goal you need to know where the potholes are. You need a vivid understanding of what might go wrong. There is an old saying in systems science, "You only learn from negative feedback." You learn from the shock of the unexpected, from knowing that you did not know. So long as things go right and our expectations are met, we do not question our assumptions. We tend to repeat customary ways of thinking and doing. It is only when nature says "No!" emphatically, that we are alerted to misconceptions and discover that we have generalized too far.

Errors are also essential to *progressive improvements*. What we decide to call an "error" may depend on the degree of quality we seek. If 30 percent of everything we do is called an "error," we keep trying to improve. If 0.3 percent is so called we may become complacent and relax. When a spacecraft is operating in a very hostile environment, you never know if "tough" is tough enough. It is probably wise to prepare for the worst, in which case improvements are endless. Earle Huckins insisted that the spacecraft deal with the "flick of the eye" problem – instructing the computer to fix a short in the power system within 63 milliseconds, the time it takes to flick your eyes – as if the system would not already fix itself after one millisecond anyway.

> We put all of what we do through a very fine sieve to test and catch the mistakes. We test things above and beyond their expected operation and temperature exposure to find interactions in the system we never designed in there but that in fact do exist. It is trust in the people and then testing and analysis to drive out and expose the bad things.
>
> Chris Jones, Cassini *spacecraft development manager*

Errors are also essential to *creativity* and to *cost containment*. What is vital is that the component or sub-assembly does the job you designed it to do. If it does this more cheaply and more effectively than before, you are way ahead. But any novel design may require several trials and corrections before it out-performs its predecessor. Banish error, and inventiveness dries up.

This must have been what John Casani meant when he told us he was "tolerant of mistakes, but not of mistaken processes." The process of detecting errors and putting these right must work properly. You correct errors and learn until no serious errors remain. Errors are what concern people the most, and they must "welcome" them in the sense of keeping their eyes peeled. Dick Spehalski, whom Goldin was holding responsible for the whole project, spent most of his time error hunting. We were told, "Spe got bored with the rock-like mechanical engineering of the spacecraft and took more and more interest in the risky areas." Spe and his colleagues Tom Gavin of JPL, Gary Parker, and John Pensinger would attend review after review.[6]

Yet willingness to recognize and hunt down errors should not be mistaken for a nonchalant approach or a cavalier attitude. You flush errors out of the undergrowth to be rid of them. "You don't go to Saturn twice," David Southwood of ESA reminded us. "This simply has to be right." At US$3.3 billion a trip, at the taxpayer's expense, we saw what he meant. The whole reason for eliminating errors is that you hold before you the goal of perfection, which cannot be seized but can be approximated by successive discoveries and improvements.

Simulating reality and playing to get serious

It has been wisely said, "Experience is the best teacher but the school fees are heavy." And if the experience is like the Shuttle disasters or building the International Space Station, the fees could be considered unsupportable. But fees are effectively waived when you *simulate* or *play*. In such cases error instructs but does not traumatize. Theater is regarded as one of the hallmarks of civilization because it simulates tragedy, thereby teaching audiences not to share the fate of the protagonists. Married couples, watching Medea kill her children to spite Jason, might think twice before blighting the lives of their own children. You are taught to use your imagination instead of suffering.[7]

Much of the animal kingdom plays. The point is to learn everything you need to know in order to survive, without hurting members of your own group, yet be ready to hunt or fight.

Cassini-Huygens as a mission has had literally thousands of simulations, almost none of them dangerous or costly. "Every 'black box' has to be put through a functional test," we were told. This is because there is no time to review the interior workings of that box for which the supplier is responsible; there is only time to discover if it works as specified. In a sense the entire spacecraft and its payload is an assembly of working models, which become the real spacecraft only when integrated and combined.

Yet one or two models of what is required may not be enough, nor can these be part of the finished spacecraft if the materials used in the models are not those to be used in the craft. You can create a computer simulation of magnetic fields or wind velocities. You can mix the gases on Titan in a laboratory jar and produce the orange haze we see through telescopes. You can drop a test parachute over Italy or the Antarctic to see if it opens in our atmosphere, and hope it will do the same on Titan. You can test whether components will distort in a vacuum or be destroyed by high vibrations. Simulations may be far from perfect, but they can still be useful or suggestive.

Often it is valuable to set up a model alongside the "real thing." Enrico Flamini of the Italian Space Agency told us, "It is absolutely mandatory to set up a demonstration model, just to see the major points of what you are building." You can make experimental changes in the model without endangering the prototype. Some changes upset the balance of the whole. If a defective component needs to be isolated and replaced, demonstration models make this simpler to do. When the probe turned out to have communications problems while already underway to Saturn, ESA's Darmstadt Operations team simulated the problem endlessly on the second version of the whole *Huygens* craft, its "engineering" model.

Another obvious reason for simulation is that humankind has only been to Saturn once with *Voyager*, and knows only approximately of the hazards it might present. Conditions *not* present here on Earth must, where possible, be contrived. Atmospheric winds at Titan can be approximated in wind tunnels, radiation can be beamed, extremes of temperature induced along with atmospheric pressures. Key components must withstand these stresses. If anything breaks or loosens the whole test is conducted anew. When batteries broke apart in vibration tests these were replaced.

It is not simply that simulations are used to educate the mission. In time, the results of the mission will be used to educate scores of Earth-bound simulations. The Europeans especially stand ready to reconstruct planetary science around what *Cassini-Huygens* discovers. Michel Blanc, an interdisciplinary scientist for the mission, explained:

> We are developing micro-simulations and numerical models, which represent the whole atmosphere of Titan, its chemistry, its dynamics and the evolution of a very complex system. As soon as we get more information from the probe, we can plug this into the system and use *Cassini* as a building block for Pan-European science.

In fact, his dream to create a European planetary science network – "EuroPlaNet" – was born out of this idea, and has recently become reality, with the European Union having approved and funded the network.[8]

In short, playing with simulations has a deadly serious import, in that it can save lives and revolutionize knowledge. The trick is to make the models do the trials, discover any errors, and so protect the precious and irretrievable parts of the mission from any harm. Even this goal is elusive, as we shall see in a moment, since it requires that the modelers detect every error. But since *some* errors are likely to get through, why not make the system self-repairing?

The self-repairing mission and the Pygmalion effect

In effect the spacecraft and its attendant engineering have been designed from their inception to be like their makers: self-maintaining, self-steering, and self-repairing. It is not just that scientists and engineers, as fallible beings, expect to make mistakes and be able to correct these; the mission as a whole, its hardware and software is designed to detect any anomalies within itself and put these right, without further human guidance. Chris Jones explained:[9]

> The software has "system fault protection" programmed into it. This can detect failures or performance outside the permitted envelope. When this happens it self-tests to find the extent of the problem and then self-repairs. I helped create the software on the *Voyager* and it helped save the mission.... The amount of coding needed for fault protection in *Cassini* was under-scoped (under-estimated). We located 290 potential faults and designed protections around each one. That takes a huge amount of code! But in the end it worked beautifully, and it is still working.
>
> Of course some faults are more serious than others, so we were especially concerned about some of the most dangerous eventualities. Suppose, for example, the second burn on the *Centaur* had failed. [It did not.] Then the spacecraft could not have been lifted out of its Earth fly-by. That's a lot of plutonium going round and around, before it enters our atmosphere and burns up. But we covered that eventuality. Had the second burn failed, a spare booster would have propelled the craft into so high an orbit around the Earth that it would not have re-entered for a thousand years.

While much of the self-repair is automated, there must be an override for human decision making. This would apply to all situations not anticipated by the programmer and hence not encoded into software. In this event, live persons and intelligent systems "talk" to each other. Julie Webster tells us (see also next page):

> The spacecraft can be told, "don't do that" at any point, by a human controller on the ground. Within limits it can be re-programmed remotely. If the signal level reaching the craft falls below a stipulated strength it will "call home" to get further instructions. We have a mode called "safing" in which the spacecraft's systems shut down to conserve energy, while operating on a fraction of a watt, yet ready to be awakened when we are prepared.

> Putting it [*Cassini*] together is not the real thrill yet, exciting and important as it is. It is when you come to work now and it sends a signal from 800 million miles or so away that it has gone in safing mode, having turned to the Sun and waiting for new instructions. Spacecraft do have their own personality, and this one is a real professional. It is like a sailboat when you know which side tacks better than the other. You get a feel for the boat and it is hard to explain.
>
> *Julie Webster,* Cassini *spacecraft operations manager and test manager during the* Cassini *assembly and integration phase*

JPL engineers told us how they learned from earlier missions that there can be a very high tension among human controllers when something goes wrong. So much is at stake that people panic and can easily misdirect the spacecraft or the orbiter. If the craft receives a garbled message it cannot understand, it goes automatically into the "safing mode" and waits for controllers to get a grip on themselves. Essentially the craft helps its controllers to calm down and think clearly.

Indeed so much of their own humanity has been programmed into the spacecraft that a kind of "Pygmalion effect" occurs, a bonding between the mission and its makers. This happens because the hardware, software, and purposeful system are an extension of the scientific community itself, and carry its hopes, fears, and yearnings. Like Pygmalion, these scientists are enthralled by their own creation.[10]

Julie Webster continues:

> When the *Mars Observer* went silent on us I felt as if a close relative had died. I needed to know what had happened, where the "body" was. I remember crying also when we lost *Magellan*. It takes almost a year to recover. It's like a personal bereavement. You cannot believe its spirit is really gone.

Bill Fawcett, manager of the Science Instrument Office for the *Cassini* project team, still has difficulty getting his mind around so amazing an achievement. "I am so awed by what we have accomplished that I get tears in my eyes just thinking about it." He did show tears when we heard him say that, as did Charley Kohlhase, *Cassini* science and mission design manager, when he talked about *Cassini*'s launch. Another interviewee likened his concern with the launch to checking the toes, fingers, and other parts of a newborn baby to make sure it was normal. When *Huygens* landed at Titan, several officials kept saying that it was "like a baby being born, and needing to count its feet and fingers."

What a mission of this kind achieves is an extension of the human nervous system itself, so that it reaches literally into the heavens – except that the instruments used actually transcend the range of human senses, seeing more, hearing more, touching more. Julie remembers walking home one night after a particularly exhausting day at work and cursing her lot:

Then I looked up into the sky and I could see Venus and Jupiter, Mars and Saturn, and I realized that part of me was up there and we would soon know much more than we know now. And I realized I was the luckiest woman alive.

Wide tolerances and built-in redundancies

Here are three more paradoxes. Why, if your target is so *precise*, do you have to build in *tolerances*? Is this not a contradiction? If there is *one critical* path that leads you to your destination, why create *redundant* paths? Why *plan precisely* when you know you will have to *adjust* continually? Do not these values negate each other?

For an answer we need to distinguish ends from means. The *end* is targeted, critical, and planned, but the *means* to that end has wide tolerances, many possible paths (all but one of which are eventually redundant), and frequent adjustments to plans which are constantly renewed.

Among the greatest obstacles in the way of this mission were the extremes of temperature. The spacecraft had to fly by Venus where it is extremely hot, since it is close to the Sun, and visit Saturn, ten times as far from the Sun as the Earth, where it is brutally cold. Heat variations are best expressed in terms of relative intensity of sunlight: at Venus this is about twice as intense as on Earth while at Saturn it is about 10 percent of what we get at Earth. That is why the craft has to be thermally protected and shaded by the high-gain antenna. Some vital components had to be cycled through the extremes, to ensure that they could cope with the materials' expansion at the highest projected temperatures and its shrinkage at the lowest. After such ordeals they had to continue operating, and the instruments had to stay at room temperature for the duration of the mission.

Instruments that could not survive these extremes of temperature had to be insulated, air-conditioned, or otherwise protected, but in a way that would not prevent them from fulfilling their function. Hans Hoffman, a German astrophysicist responsible for the thermal design that would protect the instruments in the *Huygens* probe, said that never had he or his team confronted a challenge so difficult.[11] "I knew we would succeed. I did not know how. We had only weeks to find a solution. When we simulated the descent there was a catastrophe. Most of the mocked-up instruments stopped working." So complex was the solution that the authors could only grasp its outlines. Hoffman's team applied *protective* insulating material, but with an *opening* so that the observations could be made while allowing for the dynamics of the material during the significant temperature shifts. The endurance of the assembly then had to be verified and validated.

Tom Gavin explained that from the very beginning the spacecraft had to be as robust as possible, with wide margins of tolerance against hostile environments. He was concerned lest scientists run out of mass, power, data storage, or data transmission capacity. An approach was also to test key components to destruction. How much vibration, radiation, dust, heat, and cold could they withstand? "We burn the hardware for 144 hours," he explained, to discover when and if it fails. The further you can push *beyond* expected limits, the *safer* you are. It takes the widest tolerances *on the way* to your target, to hit that target *precisely*.

The same paradox (of ends versus means) applies to having many *redundant* paths in order to find at least *one critical* path to your goal. Jean-Pierre Lebreton, the *Huygens* science and now also overall project leader, explained that they had built in many redundancies into the probe so that if one system failed the second would be triggered into action:

> There is a less than 5.0 percent chance this will fail and that we will not hear from the probe again once it descends at Titan. There are several redundancies but these cannot be everywhere; for instance the heat shield has to work, the parachute has to open, the antennae must receive and transmit.

Others were less sanguine. They were conscious that by the time the probe reached its target, the parachutes would have been in their containers for almost ten years. Titan's atmosphere has far more pressure than the Earth. Would all three open at −292 °F (−180 °C) or more? In the end, they handsomely did. And, yet, if Hamid and the probe development team had not built a redundant communications channel between *Cassini* and *Huygens*, there would not have been a successful Titan mission, as one of the channels did not turn on when it was supposed to, probably as a result of a human programming error.[12]

Chris Jones added optimistically:

> We have so much redundancy on *Cassini* that no single failure can result in the loss of the mission (although serial failures could). We have reaction wheels to stabilize the craft. There are three axes so that there are three wheels that can turn. Yet for back-up we have a spare wheel that will replace any one wheel that malfunctions – as in fact happened during the cruise to Saturn. The spare wheel is on an actuator, which can change it over. This was another creative idea that saved loads of money, time, and mass, while providing the required back-up.

Large parts of the human brain are redundant, yet "lie in waiting" to be used if injury occurs or massive challenges confront individuals; or perhaps they will come into general use if and when the human race develops the ability for greater complexity of thought. We are purposive beings, and in order to achieve our purposes we devise alternative paths to our chosen destinations. There is no one critical path, but many. What is critical is getting there, and redundancies make sure we do.

An example of a related paradox here is planning and adjustment, an issue raised by Hans Hoffman. He frequently had to adjust his *Huygens* integration plans. Few components or sub-systems were delivered just when they were needed for assembly, and often components would not test out positively at the first try. Every day brought a new surprise and unexpected adjustment. And yet he laid careful plans, as Germans so often do. Why plan if you know full well that unforeseen events will change these? This goes to the heart of dilemma theory, which is all about difference. Unless you *have* a firm set of expectations, however provisional, *you will not notice the events that upset them.* You have to have assumptions to learn that these have been confounded, which is why you revise plans every few days before adjusting to changed realities. You must discover you are off course, like a cybernetic helmsman adjusting his steering as he travels through turbulent water (de Geus 1988).

Openness to supportive criticism

It was NASA and the *Apollo* mission that first invented the project group organization. This consists of a flat pyramid of small, face-to-face problem-solving teams. Each delivers its solution to the integrative team above it, and so on up, until the mission is ready to launch. Each team has the diversity of expertise necessary to forge its solution. Each leader has a span of control over team members of a size that permits intimate knowledge of them all.

There are two distinctive characteristics of these teams, which are highly relevant to the speedy correction of errors and the process of accelerated learning. The teams and their leaders *support those who have made mistakes*, and are able to be *super-critical of colleagues while remaining friends*.

The words "mistake" and "criticism" are not used here in a negative sense, but as elements of learning and communications within and among the teams. The whole culture could not work but for subtlety of the human organization.

Dick Spehalski had co-located all the units on one site in the JPL building so all project team members could talk to one another; these units consisted of face-to-face teams of close colleagues. It created a network of problem-solving teams, small enough to be intimate, yet part of something as important as keeping a very critical eye on potential errors. Where else do people who love and admire you, also criticize you and encourage you to improve? In close-knit, nurturing families criticism is usually constructive. Parents and siblings say in effect, "I love *you* but this particular behavior must cease."

For Chris Jones it was all about being supercritical of those you admired and supported. Of course, they were good, but were they good *enough?* This mission was *so* challenging, trying to do things that had never been done before. It required extraordinary capacities.

JPL and its culture consist of "not being satisfied with the status quo." When we find something we have done wrong, that is like nutrition for us, and we applaud ourselves for having found it. HQ may not see it that way, but we feel, "I am here to find fault." It looks like we are all enemies but we dissect a person's work until we find a fault. That is our joy and the value of our periodic reviews. Some consultants we hired thought we went too far in this, but our find-the-fault culture modifies behavior very effectively.

Spe says it even more succinctly: "I value the absence of error."

Here is what the Italian Bernardo Patti, a thermal/mechanics engineering manager on the *Huygens* development team, had to tell us about the team:

We knew each other so well that we engaged in "positive discrimination," making fun of the others' national characteristics and teasing them about those. Because everyone felt accepted no one took offence. The mild teasing relieved tension and made everyone laugh. To have all Italians would have been boring. We are all full of prejudices, but at least we can laugh at ourselves and find out how wrong we were. It was a wonderful atmosphere. I learned so much, especially from Kai Clausen and Hamid.

> Michel Verdant's favorite cross-cultural quips were about the Dutch and the French, since he had spent many years working in the Netherlands at ESTEC: "The French restaurants have stopped serving cheese platters to the Dutch as they eat all the [free] bread with it and that's it." We also liked his comment that he has no problems with the Americans, because "the French and the Americans are very similar."

"Hamid would always support you even if and when you made a mistake," Michael Verdant, payload and instrument manager on the *Huygens* development team, told us. It was important for everyone to appreciate that it was not the mistake that was supported, but the person. It makes sense to support the person for many good reasons: so that he or she will accept being wrong; so that others will admit to being wrong too; so that more mistakes will surface and be corrected; because it is only when we acknowledge error that we learn how to improve; because management wants that person to persevere; because that error may be the clue to many more, and the individual might know where these are hiding; and because near-perfect missions are the work of fallible people, checking up on one another.

"In all our efforts," explained Kai Clausen, "we never pointed fingers. We regarded mistakes as normal and inevitable. It was never in our tradition to find anyone 'guilty.'" Another interviewee observed:

The priority is to solve the problem. There is not much time. We do not cast blame or search for culprits. To do so is to think backwards. We assume it's an honest mistake and go from there to a solution from which we can all benefit.

The interesting aspect is the total absence of malice. Errors are precious commodities, like nuggets of gold. Yet it is not always easy to criticize because you can get swept up in the general momentum. When you are seeking launch approval and your critics are trying to stop you, it can be very difficult to take their side, as it were.

Hence, stellar achievement requires that we sustain people but critique their efforts and expose their errors in a way they find constructive and instructive. Every error is a stepping-stone to the stars. We hold people in bonds of affection yet reform their conduct with serial objections.

Before we draw unjustified conclusions about the general forgiveness of all errors, it is important to make a clear distinction between different kinds of error. The deliberate breaking of known and accepted rules, as when a pharmacist consciously misprescribes a drug, may need to be penalized, especially when the consequences are dire. But errors made in *discovering* the rules are a way of inquiring of nature herself. An error may contribute to the discovery of laws, just as confirmations do. Nature says "no" to us more often than she says "yes," as the pioneers of aviation well know. To penalize those who failed before you is to undermine the foundation of your own knowledge. To err in trying to discover laws is not to defy these laws, but to search diligently for that by which we are all bound.

Scrutiny and trust

One issue on which our interviewees were divided was how well they trusted one another. Some marveled at "the climate of trust." People would go down dark alleyways together. There were the bonds and nostalgia you find among comrades in arms. Others pointed out that everyone scrutinizes everyone, and trusting them not to err is what you must *never* do. Here again is a paradox. People and their actions were indeed scrutinized very thoroughly, and *because* this was done, the reliability and trustworthiness of individuals and teams became increasingly evident, as did their good intentions. It seems clear that most errors were inadvertent. An intriguing example is outlined in the box on page 63, where JPL staff had to solve the gigantic scheduling challenges between teams and with contractors, and so on. They invented a system with the sophisticated name of "RecDel."

We described earlier how many told us that *Cassini-Huygens* was the highlight of their professional career. The brainstorming was wonderful. The discoveries were exciting. It was only when you repeatedly question each other and still the basic proposition survives, that you realize that here is a person you can rely upon, who has been tested to the hilt, yet remained true. Team members found that others often voluntarily, and at trouble to themselves, came to their aid. It was not at all uncommon for one unit or institution to solve a problem that arose in another.

People will also behave in trustworthy ways if they know someone else will be scrutinizing them. They will forestall external criticism and examination by self-criticism and self-examination. We learn to trust people over time and if we have not scrutinized them carefully, how can we know to trust them? *Cassini-Huygens* had high levels of trust born of careful scrutiny.

There is a second source of trust: the assumption that people have high capacities and good intentions. Here the trust comes from not being able to second-guess an expert but trusting him or her to solve a problem. Hans Hoffmann had no idea how he could solve his thermal design, but he trusted in the genius of his team members that a solution would be found. *Cassini-Huygens* is an assembly of expert systems, and it was important for participants to trust the capabilities of experts in other fields. You can not tell them what to do.

> Well, what comes first? The chicken or the egg? Maybe this mission is going well because of the people we have on the project, or the people relate well … [because we have built trust]. At a certain point you will be in a position in which you, your work will not be black or white but with a shade of gray. Then you need to have full harmony to solve the challenges as we for example had with the antenna's paint cover, and we did solve it together, JPL and Alenia engineers.
>
> *Enrico Flamini,* Cassini-Huygens *project manager, ASI*

Qualities emerging from paradox

We have argued here that perfection can only be approximated by an *accelerated race to learn*. Human beings are highly fallible but their errors, if admitted and exposed, can be corrected and eliminated. Moreover, through erring we learn faster and more vividly.

This process is circular and improves continuously, reconciling paradoxes and dilemmas along the way. *Cassini-Huygens* was designed, reviewed, and redesigned. The team pounced on errors and corrected them. It used a considerable number of simulations to err without serious costs, and to play seriously and so learn. These feedback loops were not simply in the social system, but in the software, hardware, and guidance of the spacecraft itself. The mission was self-repairing and even "comforted" its own controllers, who in some cases fell in love with the "humanity" of their own creation, as did Pygmalion.

The spacecraft was targeted towards a critical objective, by careful plans. Yet wide tolerances were necessary to get it there, and the path it eventually chose would render other paths redundant. Every plan would need later adjustment. All this can only be accomplished if teams have members who are intimate with each other and have social bonds affectionate enough to withstand strong criticism and fault-finding. Such corrected faults pave the path to near-perfection. You come to trust each other because you have checked up so carefully. Figure 3.1 takes just two of our dimensions, *designing* and *reviewing*, playing with *errors* and seriously *correcting* these, and makes of these an illustrated learning loop. Mission members learned *from* playful errors how to correct these seriously. We can also express this in multiple dimensions – see Figure 3.2 (overleaf).

The other value contrasts featured in this chapter can also be expressed as learning loops, thus *supporting* team members yet being *critical* of their work, *scrutinizing* them minutely, so learning to *trust* them, making the mission *tolerant* of extremes so that it will be *on target*. These dynamics are not exclusive to space missions, but apply to all enterprises dealing with complexity and positioned at the leading edge of knowledge in their field.

Any company or organization that can unleash the kind of energy as is stored in this mission will be great indeed. And it will be even greater if it knows how to reconcile the values of many incomparable individuals fulfilling their common mission. That is the subject of the next chapter.

Figure 3.1 Error and correction

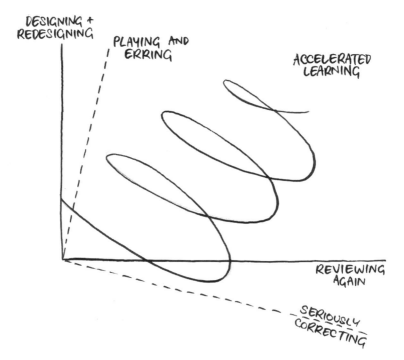

Figure 3.2 Accelerated learning

A homemade customer-driven scheduling solution: JPL's "RecDel"

Programs for detecting and wrecking errors are typically systematized. Such was the case with RecDel, which stands for "receivables and deliverables." The schedule for constructing the spacecraft and getting it ready for launch is a formidable undertaking. Literally thousands of sub-assemblies and components must converge on time, to be integrated into an ever-enlarging system. JPL's *Cassini* team had intended to use Primavera, an off-the-shelf computerized cost and schedule control tool favored by the Department of Defense and used by numerous US government contractors. Alas, it is of daunting complexity, requiring over 25 employees to run it successfully. These are experts in scheduling who are often largely ignorant of what has to be scheduled and why. Much time of the project team is wasted educating schedulers, so it is no wonder that Primavera is sometimes called "Big Brother."

Not for the first time in this project, seemingly insurmountable hurdles triggered a creative response. Michael W. Hughes and Reed E. Wilcox created and patented the RecDel approach under JPL's name. The *Cassini* project team's scheduling challenge focused on what had been agreed between parties about delivery and receipts. What were the agreed delivery dates, and were both sides confident that they would be met? Where "delivery" of a kind had already taken place, did this comprise the wanted solution, or was the recipient still unsatisfied? The "customer" (or recipient) had to be satisfied about promised delivery in order for a contract to be in force. The recipient also had to be satisfied with what had been delivered for that contract to be seen as complete.

Unlike Primavera, RecDel took no note of the internal schedule of any unit. This was its own business. Indeed if units were to deliver on time and thus meet their external promises, their internal schedules needed to be kept flexible and changed to meet urgent requirements, as when extra persons might be assigned to a project that was running late. What was monitored were "promises kept" or otherwise, not *how* these were kept, which might have required emergency action. This simplification saved 44 work years and reduced the number of staff needed to monitor the information system to six.

The focus of RecDel was interfaces among people, and the system encouraged face-to-face meetings, and if necessary the renegotiation of specifics and deadlines during such meetings. Unlike Primavera, it was not an instrumental way of inducing compliance, but a record of agreements made and commitments met. It was a record of reconciliations, a term that plays a major part in this book. It covered verbal reconciliations and technological reconciliations, and specified that such reconciliations should be the obligations of both parties, which needed to work through their differences and reach agreement.

The system did not take sides in any disputes, by upholding historical understandings. Instead it alerted the whole organization to relationships that were not working and needed to be made to work.

Senior managers constructed a "most wanted" list of typically between 10 and 20 projects where agreement on terms or on deliveries had not yet been reached. In the *Cassini* development project, projects that remained on this list came to form a "Maggot list," and finally a "Cadaver list" for relationships considered dead. These were fortunately very few, and few triggered intervention or removal of the principals. Those responsible for listed projects had to appear personally to answer questions at the Monday staff meetings, and failure to agree with partners exposed them to a degree of shame, which intensified in proportion to the delay. It remains important not to blame just one party to the interface. The issues are far too important to subordinate one judgment to another. It is reasonable to blame both parties for a breakdown in communications that threatens a project's survival.

RecDel proved extremely popular. Managers perplexed by Primavera found it easy to understand. While the latter was largely a technical system, RecDel was a socio-technical system of the kind long advocated by, for example, the Tavistock Institute in London, UK. What it monitored were social agreements about technical issues. It was up to each expert to fight fiercely for what s/he believed in, while not jeopardizing the mission with extended quarrels.

The sub-system managers were especially enthusiastic. Their concerns could not be overridden without this being registered within the system, and this gave them a chance to explain new solutions. Unlike Primavera, the system pushed decision making down to the interfaces where the serious work was done. It actively encouraged parties to talk to each other, and this was made easier when teams were colocated. RecDel was well adapted to a project where uncertainty was rife and changes were frequent. It was part of the corporate culture, instead of militating against it. It changed when understandings changed, and registered new ingenious solutions. There were as many alternate paths as there were new agreements. Above all, it flagged extended disagreements as "errors" and "corrected" these.

The system spread to include ATLO (Assembly, Test, and Launch Operations), and when the spacecraft finally roared into the sky above Cape Canaveral, it was being lifted by thousands of monitored agreements by distinguished engineers.

4

Competing to cooperate: how thousands of incomparable individuals fulfilled their common mission

> *When a team outgrows individual performance and learns team confidence, excellence becomes a reality.*
> Joe Paterno, legendary Penn State football coach

The near-worship of competition by Anglo-Saxon-speaking cultures often misses the wider context in which these rivalries occur. To the extent that these are non-lethal, non-injurious forms of *play*, the whole enterprise and community can benefit. Having colorful protagonists assert themselves and their disciplines enables the larger enterprise to learn from the best of their propositions. But these games are less finite than infinite, as we all share the genius of winning contributions while the enterprise reorganizes itself around the best initiatives. It is the knowledge arising from playful rivalry that continues to the end of time, and informs human cooperation through shared truths.

Yet in the folklore of Anglo-Saxon capitalism, everyone is assumed to calculate his or her self-interest and pursue it. The only way to control this

Leviathan of struggling rivals is to make them subject to a common authority, the crown or the constitution. Without this, the war of all against all would ensue and our lives would be "nasty, brutish, and short." Margaret Thatcher famously doubted whether "society" could be said to exist. We were rather an aggregate of divinely discontented individualists prepared to abide by "the rules of the game." Our laws are "the harness of self-interest," as American economists like to put it.

Were this the whole truth, *Cassini-Huygens* would not now be circling Saturn, beaming back to Earth information that doubles or redoubles our knowledge almost daily. It is the theme of this chapter that indeed *Cassini-Huygens* was and is a hotbed of competitive rivalries, but it is *also* an extraordinary feat of seamless human cooperation, one that deserves the status of a new wonder of our world or solar system. All this has been achieved without the help of those contemporary objects of secular worship, the profit motive, or the identification and defeat of some rival. Rather they were the achievements of the international space engineering and science community, located in many universities, laboratories, company facilities, institutes, and government agencies. Monetary rewards were modest, at least for most.

So far from competition and cooperation being enemies, as the cold-war warriors tried to convince us, these can be complementarities, as we shall see. It is because a person is fiercely dedicated to his or her own cause that a composite of such causes can perform superlatively, and a greater truth can be unveiled. While it is true that Western ideology once organized itself around competition, while the Communist bloc organized itself around cooperation, the net result was mutually assured destruction, aptly known as MAD. Historians may one day conclude that it was bitterly ironic that we threatened to kill each other over what were a demonstrably false dichotomy and a form of pernicious dualism. One of the important lessons *Cassini-Huygens* can teach us is how these contentious values were reconciled. We will proceed in the sequence that follows, letting particularly the scientists this time amplify our argument.

1 In the wake of the Argonauts: a boat full of princes and captains.
2 Furthering the ideal of Europe.
3 Dangers and delights of diversity.
4 Science as joint discovery.
5 Heavenly sequences or "take your turn."
6 Twelve ways of sensing: interdisciplinary science.
7 Cohesion, relaxation, and informality.
8 Orchestrating the uniquely excellent.

Weaving these together, each principal was like an Argonaut in the same boat, many inspired by the ideal of Europe and bringing the diverse perspectives to a delightful inclusiveness. They agreed to be disciplined and joined by scientific discovery, to "take their turn" in celestial orbits, and observe one complex system in different interdisciplinary ways. They learned to relax and party together, and were managed in ways which orchestrated excellence.

In the wake of the Argonauts: a boat full of princes and captains

Toby Owen chose this metaphor to illustrate the mission:

> *Cassini-Huygens* is our quest for the Golden Fleece. We are the modern equivalent of the Argonauts. I chose Jason and the Argonauts as a source of names for features on Saturn's moon Phoebe because the trip of the *Argo* resembles that of the *Cassini-Huygens*: Saturn was at the edge of our solar system to the ancient Greeks, and Phoebe is at the edge of the Saturnine system just as the destination of the *Argo* was at the edge of the Greek's known world. Beyond that, this comparison makes sense in that many Greek city states and others in the Mediterranean region chose their local heroes, their captains and princes, who won a place in the *Argo* and agreed to face the hazards of their journey in return for a share of what was discovered.
>
> They sailed through the Dardanelles, from the Mediterranean to the Black Sea, opening up new trading routes to cities of fabled wealth, especially Colchis.
>
> In much the same way we have collected "princes" of their own disciplines, persons who are literally the best in the world at what they do, and offered them a share in our exciting journey into the unknown. We are all highly specialized yet we share the same fate, the same reliance on one another, the same burning curiosity, the same thirst for knowledge.

One characteristic of scientific specialization is that your own piece of the giant jigsaw puzzle wins *all* your attention. In a mission lasting more than a quarter of a century, this piece is the legacy of your working lifetime. Nothing is more important than proving or disproving the theory of gravitational waves, or of finding a methane ocean beneath the crust of Titan, or discovering volcanic activity on icy satellites. Anyone who imagines that planetary scientists did not compete to be included in this mission, and once included did not fight for every precious minute of observation and experiment, is deluded. All had responded to an AO, "announcement of opportunity." All had written proposals, and all had agonized while the decision was made to give them the biggest opportunity of their lives. The excitement was electrifying, the stakes huge, the rivalry intense, the tension palpable. After all, it has taken seven years to make this journey, and twice as long to prepare for it. Now their time has come.

This, then, is a mission of hundreds of chiefs and relatively few Indians. If exploration and discovery are to be maximized, then all these chiefs need to have their dreams fulfilled. Nor is there anyone "above" them with enough knowledge to overrule this specialist or that one. Each is the sovereign of his/her own domain, an ego clamoring for recognition – and rightly so.

Planetary science is a complex latticework of connections. All specialists see *their* nodes on this net as centers of meaningful connection stretching outwards, much as one piece of a jigsaw contains clues to the pieces that surround it. Each piece can legitimately be regarded as a lynchpin organizing the wider structure. We can safely assert that never was there *competition* more intense than that among engineers and scientists on this mission. Very rarely in life do a group of people pry open a treasure chest of knowledge this large, or get to satisfy a curiosity burning since the age of Ptolemy. Can we wonder that ambitions soared?

> I had to write a 15-page reply to the AO. The last 45 days before the deadline I spent seven days a week on this outline, and put everything else out of my life. I was very focused and motivated, motivated because I am a scientist, I do research, that is my habit. I am a junkie, and want to do research and publish; I am addicted to it!
>
> Robert Nelson, JPL scientist and member VIMS science team

And yet it is difficult to imagine a mission in which *cooperating* smoothly and comprehending deeply the needs of others were more vital to overall success. This was a project with more than a thousand integrated sub-systems, any one of which could impair the overall mission either slightly or catastrophically. Never have so many ambitious individualists been more dependent on the expert professionalism of colleagues, with not just their own destinies in their hands, but that of everyone. Individual responsibility really comes into its own when you accept responsibility for the larger system integrated with your own. Ten pairs of eyes searching out mistakes in your particular contribution are better than one, and you do the same for those adjoining you. You are responsible for your piece, for the seamless connection of your piece to other pieces, and for any fault in those other pieces that would damage your own. If the pressure to *compete* is intense, so is the pressure to *cooperate*, and the need to fine-tune these contrasting values.

What a venture such as this one demonstrates is that it is possible to be both more competitive *and* cooperative: that competition allows the best talents to emerge, while cooperation allows a mission to self-organize around them. It is not easy to optimize these values, but we believe it was achieved in this case. Any organization needs to know what its people most excel at, so that it can arrange for the best to help and sustain the rest, so that, as the challenge shifts, the best qualified, once again, take the lead. Management scholars call this reconciliation "co-opetition" (Brandenburger and Nalebuff 1996).

Furthering the ideal of Europe

Those participating on the European side in this project were committed to the ideal of European unity, to including the United States in this quest, and of spreading the idea of cooperative endeavor to a world teetering on the brink of clashing civilizations and wracked by new schisms.

Daniel Gautier did not mince his words:

> The only way to build a cooperating world society is to bring this stupid [national] rivalry to an end and to explore together. We now seem to be going our own way again in ESA and NASA and such competition is a regression. Together and by sharing what we know, so much more could be accomplished. Science changes the lives of people in largely benign ways, especially when we share the outcomes. It is the measure of our humanity. When de Gaulle championed space exploration in France from 1958 to 1969, it changed everything for us. Yet since 1990 funding from the government has declined every year. In 1990 it was 1.15 per cent of GDP. In 2003 it was 0.85 per cent. Will France turn its back on science? Not if we begin to see that science is essentially international and interdisciplinary, that it's a way of respecting our fellow human beings and that no nation can afford its own space program.
>
> I believe in the European mission towards a peaceful world. Do you realize there has been no major war in Europe since 1945? That's nearly 60 years, the longest interval of peace since Roman times, brought about by peaceful cooperation. The difference between the United States and Continental Europe is that we have seen continuous wars on our territories and they have not. Claude Monet was the Father of Europe when he founded the Coal and Steel Community, which became the Common Market. The Second World War was the great trauma that brought much of Europe to its knees. We were in a terrible state, but we have learned that thinking and discovering together is the antidote to war. Unlike the United States, we are a mosaic of cultures in dialogue – not a melting pot. I believe *Cassini-Huygens* is a testament to that idea. Creative and imaginative people must be given a mission in life if they are not to turn upon each other.

We have quoted Gautier at length because his passion infuses much of the European contribution to this mission. Following the collapse of the Soviet Union, France is one of the few European nations where the ideal of a cooperating world remains alive and the sovereignty of unalloyed individualism is questioned. It is important to grasp that when dealing with ESA or ESTEC, nationalism has already ceded place to Europeanism. Toby Owen told us:

> I remember the first time I visited the ESTEC facility. People from many different countries were all working with each other, not because they

were afraid of some third party, but for science and shared curiosity. I was very moved. If we could get scientists from across nations to join and share this sense of excitement and adventure, we might not end up with pre-emptive wars, but rather engage them in cooperative and instructive work of which their nations could be proud.

What I believe we have created is a beautiful demonstration of what human beings can do when working together. The wonderful thing about space is that it infringes on no one's turf. Instead of bombing schools, roads, and hospitals, why can't we join in exploring *terra incognita*? You cannot long explore space without realizing the fragility of our own planet, this precarious balance that we destabilize at our own peril. Global warming can be seen as a reality. Venus has a greenhouse effect that makes life there impossible. What we see raises real questions about the future of our Earth. *Cassini-Huygens* is a possible answer to our predicament. It shows what thousands of people from different cultures can accomplish, when working together on something vital to us all.

When you hear the passion towards the world from Daniel and Toby and consider the success of this planetary mission, you wonder why our leaders continue to propose unilateral space ventures. Did G. W. Bush miss a huge opportunity by not proposing that a human mission to Mars should perhaps be a truly global effort, not led unilaterally by the United States? Should he have known more about *Cassini-Huygens* and its strength in diversity?

Dangers and delights of diversity

Diversity is commonly seen as a problem, the trigger for genocide by Turks on Armenians, Nazis on Jews, and Hutus on Tutsis. Differences among cultures arouse anxiety, and anxiety can make people crazed with hatred. They blame outsiders for their own accelerated heartbeat and respiration, for the sweat on their skin. Outsiders are claimed to be subverting their confidence, undermining their powers, as, quite by accident, they sometimes actually are.

But while stress pushed to an extreme can make us anxious, stress in more moderate degrees makes us excited, curious, and enthusiastic. Tourists seek the excitements of novelty, the spice of difference. They enjoy variety. Studies of how rival groups can overcome their differences and enjoy each other have made much of superordinate goals.

It is a well-known axiom of social psychology that if a challenge is big enough while still feasible, all enmities will be dwarfed by the thrill of the common challenge. In this process differences between people suddenly become a source of enjoyment and admiration. How else can we explain the testimony below? For example, Kai Clausen, a German working for ESA, told us:

> It's all so colorful and enjoyable. It's not like working for an all-German company at all! I guess any European venture pre-selects open-minded people who want to share and learn. The cultural differences are a delight, but we scramble over these hurdles because we have one common objective.

Al Diaz, an American until recently leading NASA's Goddard Space Center and now the associate administrator of science at NASA Headquarters, explained:

> It was tremendous working with Europeans. You suddenly discovered your mutual dependence. Those at the top had created a vision, a vision of great excitement and meaning and we were actually going to *do* that.

"I'd walk together down a dark alley with those guys," Marcello Fulchignoni told us. "Together we have left not simply a legacy to science, but a powerful symbol of what different people can accomplish."

David Southwood, ESA's science leader, believes that "international science and engineering is the best because you can locate the world's very best knowhow. Russian rockets are still more reliable than ours. Italian space communications and radio science are unequalled. The wider your search, the better your team will ultimately be." This was the general consensus of those we spoke to. Marcello pointed out that values are diverse, and nations tend to do best what they most value and admire: "Different nations excel at different tasks. Consider a team that combined the Montgolfier brothers with Graf Zeppelin and the Wright brothers and you get an idea of how widely talent is spread."

What this project made possible was a wide diversity of starting points with a single culmination. It is wonderful to meet strange people provided you ultimately understand what they are saying, and come to appreciate their views. *Cassini-Huygens* was a mission in which a *diversity* of cultures came together to form a *meaningful whole*. People are fascinating because they are different. Those who are surrounded by clones and echoes of themselves will soon grow bored. "We can only enter into a relationship with one set at a distance," wrote Martin Buber, arguing for the reconciliation of distance and relationship.[1] It is *because* others are different that we learn while engaging with them, and those engagements

> I like the international aspect of our science work. You meet so many interesting people who are very, very good. As time goes by you collaborate more and become friends, even though there is much competition too. And, you know, you just become more tolerant and open to different cultures. For example, the Italians kissing everybody, like we Brazilians do too, it is just in the culture and is simply a way of showing friendship.
>
> Rosaly Lopes-Gautier, JPL investigations scientist, VIMS team

are memorable. In short, diversity is a form of risk taking in which we gamble to achieve a more exciting, creative, and informative relationship. Strange people we come to understand are all the more fascinating for their strangeness.

Science as joint discovery

Of all space missions ever attempted, this one has the most scientific significance. It was scientists who conceived it, fought for it, and saved it, supported by passionate engineers.

What does science seek, and to what higher order is it ultimately beholden? Where everyone is brilliant yet different, who or what brings order to this whole? We obtained a clue to this in listening to Dave Southwood: "Science is not political. It is supranational and beyond the nation is a unity acceptable to all, a clear goal with a definite conclusion."

Roger Bonnet believes that "science is a universal language." The French cooperated with the Soviet Union in the 1980s, and he believes these efforts were justified, since this common language is one of mutual understanding and humane discourse. The Italian planetary scientists did the same even in the 1960s.

How valid are these claims? Science disciplines the investigator into asking questions of nature and abiding by the answers. It is a common admission that we cannot be proved right, only wrong; that if we disagree with one another, nature is our arbiter. Moreover, the more questions we have and the more viewpoints we include, the richer will be the information we receive. Science is a way of civilizing discourse and enriching knowledge.

Some organizations seek profits, some implement legislation, some heal or litigate, but all these must *learn* to better meet their objectives. A science mission differs from these only in the fact that exploration, discovery, and learning are everything. Scientists are not learning to make profits, or learning to help clients, they learn for the sheer delight of knowing. Dick Spehalski recalls giving a subordinate a raise and feeling apologetic about how modest this was in relation to his talent and contributions. The young man answered, "I should be paying you guys for letting me work here [at JPL]." Many told us that this mission was their reason for being.

Heavenly sequences or "take your turn"

We have seen that the scientists and engineers agreed early on to eliminate the rotating and scan platforms on the spacecraft; the earlier design simply was too complex and unwieldy, "inoperable" as the reviewers had said. The instruments now being bolted to the *Cassini* depended on the craft turning before they were pointed properly. In other words, much more complex science observation sequences had to be developed and programmed into the orbiter's command and data sub-system. This made it far more complex to operate *Cassini* and the instruments than would otherwise have been the case. But the argument could be made that this sequencing is one of the secrets of better cooperation, taught to us in kindergarten, when the teacher says, "Take your turn."

There are a considerable number of these turns: each of *Cassini*'s 76 orbits around Saturn will have many of them. All twelve instruments and the teams managing these instruments had to learn how to cooperate on a science schedule, even as the spacecraft was on its way to Saturn. That agreement was far from easy to find. The principal investigators behind each instrument wanted to make such long and detailed observations that the time available was over-subscribed five times. Many conflicts had to be worked out in the target working teams (TWTs) established by the leadership for the various targeted science areas.

Bob Brown, team leader for the VIMS instrument, referred to the process of seeking agreement as "herding cats:"[2]

> Only pride in the work they must assemble is the glue that holds them together.... It's very competitive. You have to fight for the rights of your instrument. If you do not assert and defend your identity you will lose it in the rush. But we are under the gun. Saturn gets closer and closer. We must hammer out agreements.

Kevin Baines, a co-investigator on the VIMS team, said:

> Big egos make for poor science. But the negotiating helps. If I give up something on Orbit 5, I expect to get a concession on Orbit 6 or 7. The fixed platforms, with instruments stuck fast on the outside of the craft were, in the end, an aid to cooperation, because we had to weigh the value of one view against another and ask, what is best for *Cassini*, for planetary science, and does this instrument help my own and others to find answers? Since it takes half an hour to turn the spacecraft 90 degrees we cannot switch back and forth. With the movable platforms we had wanted originally, paradoxically, we would have learned less about each other's work. Of course there are more orbits on this mission and that helps. Even so, you hear guys say, "Seven years it took to get there, and six years of development before that, and now you're giving me four hours!"

Dennis Matson, the *Cassini* project scientist, saw sequencing as vital to the management of the scientists involved:

> The "taking turns business" has worked out very well.... We worked hard to increase the number of orbits and fly-bys so as to increase the

number of observational opportunities – so the range of opportunities is large.... Remember, we do orbits and observations in orbit groups, depending on where the craft is – icy satellites, rings, and so on – so people have periods where they are not competing for the craft's time.

Matson paid special attention to the scarce opportunities, for example when the craft passes closest to the mysterious "spokes" in the rings around Saturn.[3] This was the very best opportunity for the mass spectrometer team, and he made sure they got it.

There are two basic ways of making observations, called "prime" and "ride along." An instrument either has *prime* control of the orbiter, so that it points at exactly the spot where that instrument is aiming, or it *rides along*, piggybacking on another instrument's prime objective. This is possible because there are several different parts of the system that deserve study. One instrument may point at the rings, allowing another instrument to focus for example upon an icy satellite.

Sequencing is an important principle in conflict management, and is the basis of the humble stop light or traffic signal. You must be prepared to wait, but not for long. It substitutes prioritization for winning and losing. Everyone's turn comes around.

Twelve ways of sensing: interdisciplinary science

Another important principle in the capacity of the individual scientists to cooperate was the salience of interdisciplinary science. Toby Owen referred to the twelve instruments wielded by the instruments' principal investigators and team leaders as "ways of seeing." These would be all focused on the phenomena being studied, and would measure different properties of the system as a whole. "It isn't just that people come from different countries, although that is important, but they also come from different disciplines which have different ways of seeing, hearing, and sensing."

Toby might have added that the twelve instruments between them vastly amplify the human nervous system. They "see" and "hear" what human eyes and ears cannot, peering through dense clouds of methane gas, bouncing

> I have had the hardest time accepting the fixed platforms. I am a perfectionist and the science will not be as good as it could have been. This has taken so much human effort and we do not have the funds and people to help us; we are terribly overworked. It will be a glorious mission anyway, yet the human cost has been tremendous. But on the other hand, every cloud has a silver lining. The fact that the instruments are body-fixed to a massive machine whose orientation is maintained by high-precision gyros means that we have had, in the end, a remarkably steady platform from which to take pictures. And as a result, our images are exceptionally sharp. And who can complain about that!
>
> Carolyn Porco, team leader, imaging science subsystem (ISS)

radar and radio signals off different kinds of surface, measuring winds, gravitation and magnetic fields, analyzing dust and taking temperatures.

Beside the scientists leading twelve instrument teams conducting investigations on the *Cassini* orbiter, there are six interdisciplinary scientists whose job is to examine the larger Saturnine system, using a combination of instruments.[4] Details of these combinations are all included in their original proposals, which compete on merit, like the instrument-centered proposals of the instrument principal investigators and team leaders. They might use, for example, both multi-spectral analysis and a magnetometer to see what these twin perspectives could yield. Dennis Matson stressed the importance of systems science, and believes that interdisciplinary science helped to define the entire mission:

> It is about systems science. In a complicated system like Saturn and Titan you have to measure everything all the time, as the system keeps changing and will not be reproducible. This is not the moon. Here there are too many variables – orbital geometry, seasons, solar wind pressures, storms in the atmosphere, gravitational interactions between bodies, and so on. This larger system included both planetary science and engineering. That is the purpose of the mission and the means of getting there.
>
> Our success lay in getting the science teams in place before the spacecraft was defined, and in this way helping to define its purpose, which is not just to "reach Saturn" but to study the entire system. With the earlier Mars and Jupiter missions, engineers first designed the spacecraft, and then the remaining room and time was allocated to science. We went for a science-oriented design for the whole mission, and we achieved it. Because a more inclusive observation helps your case in bargaining for the time you want, observations tend to expand in scope. "Oh wow! Are you doing this? We were planning to do it too!"

This science team relied on the interdisciplinary scientists to mediate fights about the relative importance of different instruments and ways of seeing. They are in a good position to grasp complementarities. Which combinations of which instruments will answer the most important questions? Is ion density more important or are surface characteristics? They will typically say, "We need both and here is why." But they will also help with the length of the sequence needed to pin down both measurements, and why they need to be done in rapid sequence. It is as if they are occupying "golden thrones above the fray."

The admiration for scientists with this cross-science role was not universally shared. One instrument scientist complained to us that far from mediating disputes, they did not always turn up at meetings:

> They considered themselves to be a High Council advising Dennis Matson, but principal investigators insisted on similar access. You

would think that they are God's gift to science. Even so, when everyone is saying "My experiment is really, really important," and the others are saying, "So is mine," the interdisciplinary scientist can sometimes help.

Cohesion, relaxation, and informality

Every cooperative undertaking includes an emotional element, what psychologists call a "relaxation response," or more recently, "emotional intelligence." We do not really trust people unless we can feel relaxed in their presence. Our bodies tell us whether we like or trust people. While officially work is kept separate from play and we do not come to work to relax or enjoy ourselves, in fact the need to relax while bonding remains vital to the challenge of integrating our efforts in the workplace.

This appears to have been the conviction of Hamid Hassan, the project manager in charge of the *Huygens* team at ESTEC, who oversaw the work of numerous subcontractors across Europe. It is perhaps ironic at a time of great religious animosity and polarization that this feat of integration was the achievement of a Muslim from a Pakistani family. He was also a British citizen and an enthusiast for Scotch whisky, as we have seen. He became famous for his choice of exotic locations in Europe for mission meetings to build rapport.

Hamid had noticed early on that NASA's rules, incumbent on JPL and other US contractors, forbade its members to use taxpayers' money for their own conviviality. Americans could not entertain each other, much less foreigners, at government expense. In contrast, in Europe there are generally no such strict prohibitions, or at least ways to bend the rules a bit. Occasional relaxation was an integral part of work routines. Hamid took personal delight in entertaining Americans and their families at (European) expense. He persistently chose attractive locations in Europe for quarterly review meetings, to which Americans could travel and then vacation later, with their families if they wished (and they often did). Such locations should be worthy of the grandeur of their joint mission. One of his favorite jokes was "*Huygens* waives the rules." He liked to do party deals informally and through friends. For him work and play were inextricably entwined in friendship networks.

This view is in part cultural. The United States is a rule-oriented culture, a value which sociologists call *universalism*, with uniform rules applied to everyone and activities judged by their political and legal correctness. France, Italy, the Indian sub-continent, and Southeast Asia are far more personal and exceptional, a value called *particularism*, with every case unique and particular. Status in countries like Pakistan depends very much on powerful, particular, and privileged personalities remaking rules.

It was said that Hamid had no family. He was divorced, and his daughters were grown up and had their own homes. *Cassini-Huygens* was his family, and he saw to it that the vitally important development team bonded regularly. *Cassini-Huygens* seems to have learned from Hamid. Many of these interviews were conducted in Venice in June 2003, amid wonderful aesthetic and relaxed surroundings. The mission knew how to party! Personal friendship had been a key to the success of this program almost from the beginning.

Yet Hamid was not everyone's favorite person. He had very demanding deadlines to impose and could be abrasive. Timelines were non-negotiable because you cannot assemble a space probe with major components late or missing. Many of our interviewees credited Anne Cijsouw, Hassan's personal assistant and *Huygens* team coordinator, for providing the warmth, tact, welcome, and social grace that made the parties into such pleasant occasions. She had to coach Hamid in how to treat Western women, starting with herself (the red roses he lavished upon her with such gallantry were not quite the behavior she desired). Anne remembers crying as the rocket was successfully launched and climbed into the sky. Yet the experience was bittersweet. She was, of course, tearfully happy to see all those years of effort achieve a perfect launch at Cape Canaveral, yet her team would be breaking up now. This was the end of one important phase of the mission. Hamid already looked ill, and his daughters took him home before the celebratory party.

> "Why is the meeting in Munich just now, when it is Oktoberfest?" inquired my boss at JPL. "Ask Hamid," I answered. "I did not plan this meeting."
>
> Don Kindt, JPL's NASA–ESA liaison for Cassini-Huygens

Within a year Hamid would be dead, and a few years later Earle Huckins also died. Yet their spirits were aloft and on their way to the far end of the solar system. Organizations and missions are not held together by hard work and technical precision alone. Before the technologies can bond together, so must the people responsible for every vital component. Hamid and Earle saw that American values would not suffice. They needed leavening with social and emotional interactions and abiding passions. Formal systems have much to learn from informal humanity, and thanks to Hamid, Earle, Anne, Spe, and many others in this program, the conviviality needed never lagged.

Although Daniel Goldin, NASA's chief, had persuaded himself that he had Dick Spehalski "nailed" with his "ass on the line," if anything went wrong, this was not how Dick Spehalski, Hamid Hassan, Dennis Matson, Earle Huckins, and team leaders treated each other or their many team members. There is more about this later in the book.

Orchestrating the uniquely excellent

How do you manage and lead the interaction of hundreds of talented, independent, specialized experts, with an unrivaled knowledge of their own disciplines? To overrule them is to place in jeopardy the whole mission, for who knows more about their unique contributions than they do?

Kevin Baines contrasted *Cassini-Huygens* very favorably with the *Galileo* mission, where those in conflict with other specialists constantly petitioned their own functional heads at NASA or JPL to intervene on their behalf and demand a fixed allotment of observation time. "In the *Cassini* program they never tried this because they knew a culture of science was in command and that we respected one another."

The French participants gave democratic names to their various groups. It had begun with Roger Bonnet, who insisted from the outset that Europe and the United States "would be equal partners. There would be no colonizing approach." ESA, he told us, was a genuine partnership of equals. Each participant had a vote. "The *political* side of NASA claimed to be international. In fact, it initially wanted everything done exclusively in its own way." Yet this did *not* apply to the American planetary scientists. They respected all fellow scientists, regardless of origins. "They became fair and open, no sooner did they realize what we could do."

Jean-Pierre Lebreton created what he called the Parliament of Scientists, with a senior group he called the Senate, consisting of (*Huygens*) principal investigators. The principle of "subsidiarity" was borrowed from the European Union. This means that every decision is made at the lowest possible level at which full information is available, on the basis of science, not politics or position.

Perhaps Dennis Matson is the prime exemplar of how to manage the people side of burgeoning complexity. He came on board in 1989 and is still active. He looks and acts like the ultimate leader of a tough constituency of very smart, competing people. Soft-spoken, and always calm and confident, his adage was, "I try to avoid deciding." Twice he predicted that the Saturn orbit insertion and next the landing on Titan would be a "slam dunk." They were.

Early on he noticed the meetings of the scientists were noisy. Principal investigators and team leaders were competing for attention and were not confident of getting it. He accordingly introduced "requests for action," which recorded in black and white what people had asked for, keeping careful track of all those requests to make sure they were disposed of in a timely manner. No one's concerns were to be omitted. He set up a quasi-Supreme Court to which dissatisfied persons could turn. Hence a conflict between the Radio and Optical teams was referred to the Atmospheric Working Group, which was directly interested in what radio and optics could do to better study the atmosphere. "I have only made binding decisions three times," Matson said. "The conflict should be solved at the place where the disciplines interact."

He also helped the *Cassini* spacecraft office change the procedure whereby certain instruments would be dispensed with if there were insufficient space on the orbiter; instruments delivered on time and within the cost ceiling were guaranteed a place. Instrument team leaders and principal investigators were allowed to enter agenda items at any meeting they attended. There were to be no unannounced decisions, no "ambushes," no important issues for discussion not notified in advance. Dennis Matson used the scientific objectives for the mission in the manner of a constitution. Whose contributions most clearly enhanced the mission's overall objectives? "My interest was always that of moving the scientific agenda forward."

Matson decided quite early in their deliberations, "We would not vote on things." Votes were mere aggregates, which could steamroller the principled objections of a small minority, who might very well be right, especially about their own specialty. Hard although it was, he looked for consensus. This obliged all advocates and all objectors to explain themselves in detail. What could be done to obviate the dangers an objector foresaw? You must neither shut critics up, nor make it easy for them to say "no." Every position was scrutinized. Every conflict was worked through. What additional information was needed to settle this dispute? He introduced an "open book" rule. All documents on all issues had to be shared.

On some issues Dennis did make definitive judgments, but these invariably favored openness and sharing. Carolyn Porco, head of the imaging team, whose NASA-financed pictures of Saturn and Titan would be on television screens and in newspapers across the world, advanced the argument that these pictures would be under control of her institution. They could then only be used with her team's permission. On the contrary, Matson made sure NASA would rule that they belonged to any science group that might benefit from them, and ultimately to the governments of the participating countries and their citizens.

Even so, the success of this mission could precipitate conflicts. Who would be the heroes of the hour? The media love to personalize things. At the press conferences after arrival at Saturn, and later after the descent onto Titan, the scientists and engineers took equal turns at the press table – taking pains to demonstrate the unity they had achieved, while not shy of showing their pride in their instrument's or creation's initial observations. The unity was now so strong that individualistic tendencies could no longer be sustained, as the story on page 83 exemplifies.

The science leadership construct of this mission, the influence of the interdisciplinary scientists, and highly individualistic scientists working together and taking their turn in sequencing, all point at the unique occurrence of *highly specialized* disciplines finding each other in uncovering Saturn's *whole interdisciplinary system*, a paradox reconciled.

Qualities emerging from paradox

Superlative performance requires the fine-tuning of competing and cooperating, reconciled into "co-opetition." Stalwart individualists worthy of the

Argo were all in the same boat, sharing the same fate, fiercely independent but utterly dependent on each other. Some were inspired by the ideal of Europe; many found in the diversity they brought to a single inclusive mission, a genius in one another.

In discovering together, disciplined by the common language of science, each question and hypothesis provides a piece of the puzzle. As the orbiter makes its 76 passes around the system, investigators take their turns in elegant sequences, often riding on each other's backs. The twelve ways of sensing on the orbiter alone – and six more on the probe – allow disciplines to converge toward one multi-disciplinary systems science.

Even the emotions are educated, as Europeans treat their more abstemious American cousins in contexts of relaxation and conviviality. Here what is needed is less top-down leadership than orchestrating persons with uniquely excellent instruments, and constructing bridges of knowledge between them.

There have been many paradoxes reviewed in this chapter. Among them are *formalities* that learn from *informalities*, *range* that complemented *definition*, and *exotic strangers* with whom *familiarity* becomes a delight. For us the two most salient value tensions are *competing* and *cooperating*; and a *diversity* of views towards the *unity* of science. Figure 4.1 shows the learning loop, with each side of each paradox on an opposite pole. The multi-dimensional illustration in Figure 4.2 shows that this circle is "virtuous" and developmental.

We have spoken of the engineers and the scientists separately; how both disciplines found ways to integrate their opposing interests is the subject of the next chapter.

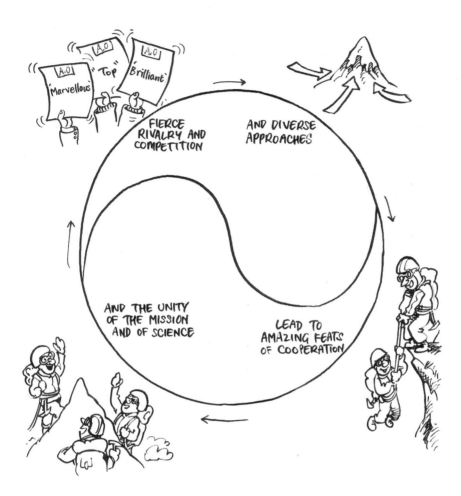

Figure 4.1 Competing to cooperate

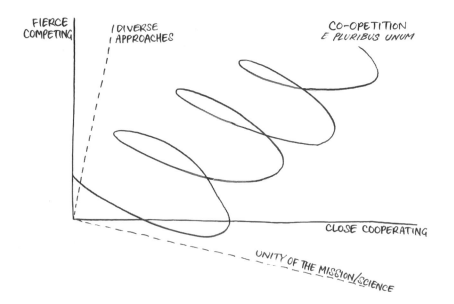

Figure 4.2 Co-opetition

Like a diamond in the sky

Twinkle, twinkle, little star, how I wonder what you are...

This nursery rhyme introduced many of us at a young age to the mysteries of the universe and our solar system. Stars are here and everywhere, and who knows, on planets orbiting some of them there might be aliens sharing the universe with us.

So it is no surprise that faraway missions to our sister planets must carry a message from humanity just in case. After all, at a minimum this satisfies our need for lasting monuments that tell our story to future generations for millennia to come. It establishes continuity in the face of our mortality.

From the early days of space exploration, we have put plaques, drawings, and markers on the spacecraft we sent, describing aspects of our civilization. In the excitement of these days of early solar system exploration, it has seemed only fitting to communicate with the beyond. The *Voyager* missions carried an old phonograph with the sounds and sights of Earth, Chuck Berry's heavy voice being one of them. Can you see those aliens rock'n'rolling on Saturn's rings? *Viking* to Mars had a little microdot with humans' signatures, just in case the Martians wanted to know who they were. Of course, it will be much more fun 30 years from now when we earthlings see this monument ourselves on Mars – when we finally arrive there, as we will.

So it was that in 1994, a small team consisting of Carolyn Porco, the Imaging instrument leader on *Cassini*, Jon Lomberg, a creative director at Time Warner, and Professor Gregory Benford of the University of California at Irvine joined forces to develop a marker for the *Cassini-Huygens*. Possible readers of the plaque would range from our distant heirs, up to some 100,000 years hence (if not aliens, who knows? Titanians?) to of course ourselves, seeing it as an appeal to our human spirit. The marker would be a polished diamond plate, to be manufactured by De Beers. JPL and ESA initially liked the idea a lot and the plan was a "go." The team actually created an ingenious design of messages depicting who we are, where we are from, how we count, and even what our diverse humankind looks like – at least partly dressed this time, as earlier naked depictions of ourselves on the *Pioneer* craft had created a furor among those who worry about such things. As Jon Lomberg reportedly said: "If we want this photo to be truly representative of all the Earth, it is no small matter to alienate a large portion of the Earthly audience."

Whatever happened, the story leaked to the press, and human emotions and claims for credit got the upper hand. The team members had a very human falling out, while NASA, among other concerns, started worrying that private industry would be major sponsors of this effort – not in the least a Japanese effort to spend US$50,000 on it, thus leveraging Japan, not a contributor to this mission in general, possibly rather easily into so costly and complex a project as *Cassini-Huygens*.

JPL had seen some of this coming, apparently, and had started its own ideas for humanity's message, initiating a drive to collect thousands of names that would go on a new marker, led by JPL'er Charley Kohlhase and aided by the US Planetary Society. ESA similarly got wind of the issues surrounding the Diamond project, and started its own signature campaign and a website to collect submissions. A favorite of many was the one sent in by a Dutchman: "If you have a job for me on Titan, please contact me at ..."

NASA canceled the diamond idea and used the pedestal on the *Cassini* to attach a DVD disc imprinted with 616,420 signatures of people from 81 countries. Mary Cassini of Australia, a descendant of the astronomer, contributed her signature. The signatures of Jean-Dominique Cassini and Christiaan Huygens were copied from seventeenth-century documents. NASA suspects the disc could last a million years or more. Encircling the disk near its edge are the flags of 28 countries that contributed the most signatures. Images of Saturn, Titan, and Earth form the backdrop of the disk. Also ESA collected over 80,000 signatures and messages over the Internet and stored them on a CD-ROM which was put on board the *Huygens* probe.

5
Engineering and science: lines and circles

Then felt I like some watcher of the skies
When a new planet swims into his ken;
Or like stout Cortez when with eagle eyes
He stared at the Pacific – and all his men
Looked at each other with a wild surmise
Silent, upon a peak in Darien.
                    John Keats (1795–1821)

"That's how we feel, it's just incredible," says David Southwood on the day after the *Huygens* landed on Titan, his lips quivering, while he cited Keats' famous poem.[1] "We don't have all the science yet, but look what is to come – 'we all look at each other in a wild surmise.'" And later: "Yesterday was the day of the engineers for getting us there, today it is the day we scientists can celebrate."

What makes *Cassini-Huygens* such a notable triumph is the elevation of scientists to a position equal to that of engineers, a form of equality hitherto rare. *Cassini-Huygens* had, from the beginning, been conceived, championed, and defended by planetary scientists. Why was there a curious subjugation of space science to engineering in other space missions? How come that the balance of influence tipped towards science at long last? Has it not been the justification for civilian space exploration all along?

That engineering thinks in straight lines, onwards and upwards, and that science is more circular and interactive – like the loops of the figures in this book – was suggested by more than one person in our conversations. Might it be possible to extend such lines so that they form parts of circles? This is what we believe was successfully accomplished in the program.

You do not have to stay long in the United States to appreciate the grip of linearity upon the public imagination. Buildings scrape the skies, trails of rocket fire shoot ever upwards, highways span a whole continent, or in Dan Goldin's own parlance, things are "faster, cheaper" – and lest we be for just one second mistaken – "better!" Here is a world in which everything escalates towards virtue, much of it the work of engineers. (The European perspective is quite different, but we explore that elsewhere.[2])

Several JPL engineers reminded us, "The rule had long been 'Engineering *über alles*.'" There are several strong reasons that engineering had dominated up to this point. Becoming skilled at sending robotic missions to our sister planets in the first place was certainly one of them. An even more powerful driver may have been the emphasis on human space flight, in which astronauts need to make a round trip and be safely returned to Earth. Given their hero status and the public's acclaim, their deaths are the most dreaded outcome of all, and only superlative engineering can protect them. What they might learn is of small consequence compared with coming back alive and enthusing the paying public with their deeds.

Once the Shuttle, a reusable vehicle, became the preferred option, sold to Congress for its ability to link space projects together and make running repairs, engineering achieved a more pivotal role than ever. If the Shuttle went down (as it did), other missions with major science objectives, like repairs to the Hubble telescope and building the Space Station, became almost impossible to fulfill.

Wes Huntress gave some trenchant evidence on October 13, 2003 to the Committee on Science of the US House of Representatives (Huntress 2003). He spoke of space exploration being encumbered by old legacies, most of them feats of engineering:

> The challenge for NASA is to throw off the yokes of the *Apollo* program legacy and to move outward beyond Earth to exotic places in the solar system, where we have been given tantalizing glimpses from our robotic space program. The Shuttle and the Space Station are the legacy of a long-past era in which the space program was a weapon in the Cold War. *Apollo* was not primarily the science or exploration program we are all fond of remembering; it was a demonstration of power and national will intended to win over the hearts and minds around the world and to demoralize the Soviets. Exploration was not what motivated Kennedy to open the public purse. Beating the Russians did. It worked …
>
> We slip the bonds of Earth not to spend 20 years in orbit studying zero-G nausea but to set foot on new worlds, learn new mysteries.… After millennia of dreaming of flight, the human race went from a standing start at *Kitty Hawk*, almost exactly 100 years ago, to the moon in 66 years. And yet for the next 34 years we have gone nowhere …

If space is a race, who would willingly carry the extra weight of a payload of science instruments? Prestige comes from hitting announced targets in highly visible feats. In comparison, the subtlety of scientific observations in dimensions beyond the human senses is barely explicable to taxpayers.

Spacecraft that get to planets and moons in our solar system are multi-purpose. They could, if necessary, hit rogue states or smash incoming meteorites. Given the conservative mood in the United States at the time of writing, "the Rock of Ages" – under American care – is deemed more important than "the age of rocks" (a comparison first made by William Jennings Bryan at the Scopes Monkey Trial in Tennessee). Engineering technology protects Europe and America. Since 9/11 this emphasis has increased. And ITAR rules now severely complicate the exchange of technical information with the very "foreigners" that made *Cassini-Huygens* so possible.[3]

There are some minor exceptions to the lower standing of science. The prospect of finding life somewhere in the universe loosens purse strings, and astrobiology is a favored calling. Finding dried-up rivers on Mars and doubtful claims to locate microbes within rocks improve funding prospects. Nevertheless many of our interviewees told us that getting science objectives to the forefront is a never-ceasing struggle. NASA's implicit attitude towards science missions has long been, "fund them, fly them, forget them," according to one of the scientists we met. He told us:

> This mission is going to give us more science than any other mission to other planets.... On the first day we get data from the *Cassini* orbit, we'll get more data than *Galileo* got from its entire mission, and this will last four or five years.... We're sipping from a fire-hose.... We'll be inundated and very little is going to be looked at by this science team, because there's not enough money.... There's going to be a major discovery, I believe, about 20 years from now, by one graduate student. On *Galileo* there are slews of orbits no one's looked at yet! What happens is you spend a lot of money making a spacecraft work right and do great stuff at the planet. You get great images ... and then NASA says, "Well, that's it. No more money to analyze." And that creates a lot of screaming in the scientific community. So now we have these data analysis programs, Jupiter-DAP, Mars-DAP, but they're all under-funded. NASA does not have the money and there's no other institution to which to apply.

We have to create perfect crafts here, that is our mission.

John Casani, JPL planetary space mission veteran

There are more obvious reasons that engineering has long had priority. It must be completed *prior* to any scientific experiments taking place. It is the job of engineers not just to get the spacecraft to Saturn, but to set it on correct trajectories through the rings, and turn it, so that different instruments can point at different objects. Data collection depends on this work being done. The scientists must wait upon the engineers. Because all science experiments are hostage to engineering preliminaries, science has for long had to make do with the time, budget, mass, and power that engineers have left over after completing their work. You cannot afford to skimp on engineering. The whole mission will be lost. You can, if necessary, squeeze the resources left over for scientists to do their work, on the grounds that some science is better than none.

As engineer Dick Spehalski explained to us:

> [In previous missions] what science was being done was determined by the capabilities of the spacecraft and was treated as a risk to the spacecraft and tended to be held at arm's length. Engineering defined what was possible and what was not possible. Science would have to take place within whatever limits were allowed. Not that this has fundamentally changed now, but the limits may be broader.

Engineering also possessed an organizational advantage. Most of its people were centralized and together at JPL in Pasadena, California, and at ESTEC in Noordwijk, the Netherlands. In contrast, scientists were dispersed across the globe in small enclaves, determined by their chosen instruments or the location of their academic institutions. Hence JPL and ESTEC had shared facilities, where most information would typically converge, and where the scientists' unquenchable thirst for data was frustrated or fulfilled. Engineering was the hub joining the sciences.

It might be thought, given these circumstances, that this would be one more mission dominated by engineering concerns, yet this was not what happened. How was it, then, that planetary science came into its own at last? What forces operated to make science a co-equal? We shall explore the answers under the following headers.

1   Differences and dialogue between engineers and scientists.
2   Europe as a partner with additional resources.
3   The crisis of restructuring: losing the platforms.
4   Setting up the trading system.
5   Science gains over time.

We begin by describing the differences in values and temperament between scientists and engineers, and showing how Europe's participation helped to raise science out of its customary second place. Mutuality was created by designing the Saturn science objectives early on into the mission and the craft's development, as well as by the growing excitement as the cruising *Cassini-Huygens* neared Jupiter for a dress rehearsal and then went onwards to Saturn. The engineers in the Operations phase of this program are much captivated by the challenge to operate the spacecraft so as to facilitate science. The curtain was going up on this stellar adventure of a lifetime.

Differences and dialogue between engineers and scientists

We were delighted by how perceptive many on the *Cassini-Huygens* project we interviewed were about the temperamental differences between engineers and scientists. We decided to capture their observations and convey how key project scientists and engineers describe each other (see the box, "Engineers and scientists: some contrasts" on page 90). So much of this could easily be projected on the differences between other disciplines as well.

There are constant themes underlying these differences. Engineers have a more linear drive towards mastery and perfection. They need to try to control every variable. Scientists have a circular pattern of investigation which rejoices in the fact that they know almost nothing, and are bound to be wrong, which makes it all the more exciting. Engineers need order. Scientists are intrigued and challenged by seeming disorder: there is so much for them to reorder! Engineers seek to conserve the capacity of their precious hardware. Scientists, knowing that it will be junked anyway, want to hazard it all, for one ultimate question. Let it expire while doing its job! Engineers are worried by too much talk and speculation. It either works or it does not. Do it and find out. Scientists talk endlessly of possibilities, of standing on the threshold of amazing discoveries. Both groups tend to see the other as a bit of a burden that their group must carry. Engineers spend the lion's share of all the money getting the spacecraft there. Scientists look on impatiently, complain, and are utterly insatiable. So many questions, such complexity, give me more!

> First you have to learn how to work with the engineers so that you read the data from the instruments the right way. It is not obvious. We had to work closely with the JPL engineers to calibrate the instrument, and we worked together as real colleagues. In the beginning they were really worried about us touching the instrument; yet they gained confidence in us and when it was the Fourth of July weekend, they left us alone working on the instrument – they had started to trust us.
>
> *Angioletta Coradini, Italian member of VIMS instrument team, Instituto di Astrofisica Spaziale del CNR*

But there was much humor and insight behind these observations. Certainly tensions were high. After all, so much was at stake. If the craft failed, everyone would have to say goodbye to the biggest explosion of knowledge in the history of planetary science, and whose fault would that be? Dick Spehalski, every bit an engineer, described this mostly benign tension well: "most of us in development enjoy the way scientists explore, but they do not always appreciate how and what got them there."

Even so, he had to admit that for many onlookers it was already over following the launch. He recalled briefing a journalist and having to convince him that the launch was the beginning, not the end. "We've barely scratched the surface," he insisted, sounding very much like a scientist. There were still amazing levels of ignorance and incomprehension between different groups.

Engineers and scientists: some contrasts

Engineers	Scientists
More linear and purposeful	More circular and error-correcting
Keep to their task	Follow loose threads
Concerned with means	Concerned with ends
Mastery and knowing	Curiosity and not-knowing
Believe they have nearly finished	Believe they have scratched the surface
Low risks with a fragile, crucial spacecraft	High risks to get once-in-a-lifetime answers
Job ends when the craft orbits Saturn	Job starts when the craft orbits Saturn
Science as a burden on engineering	Engineering as a burden on science
"You do want the spacecraft to get there, don't you?"	"Only if it can do the science."
Conserve the capabilities of the hardware	Exploit its capabilities to the maximum
There are many, many opportunities to observe	Never let an opportunity pass by
If it's working don't mess with it	Make it do more for me!
Engineers are pragmatic, cut the talk, "Do it!"	Scientists are speculative, talkative, "What's out there?"
Resemble builders or developers	Resemble artists or explorers
Enjoy rules and pre-calculation	Change rules and revise
Results oriented, want it to work right	Curiosity oriented, "What else can it do?"

Inevitably engineers cannot know everything about the science to which their work is leading. Nor can scientists appreciate sufficiently what engineers do to facilitate their operations. Candy Hansen, JPL investigation scientist and team leader of the Titan orbiter science team (TOST),[4] made it her special mission to listen to the communication between scientists and engineers on her team and reinterpret what each said to the other. She soon put her finger on the roots of most miscommunication, as neither side would admit to ignorance about what the other side was asking for.

"Engineers would often say, 'It's not in the requirements,' or 'It's too risky.'" Candy translated that to mean, "I don't really understand what you're asking for. I wouldn't know how to get started, so I'll first say, 'No, it can't be done.'" She learned to "listen between the lines." If the engineer said, "We can't do that because of the thermal constraints," safe in the knowledge that the scientists would not know what those were, Candy would make sure the engineer really understood what was wanted. "Could you give us the right ascension and declination of what you need?" "Suppose he was to give you the coordinate of the interstellar medium, would that help?" After questions like these, the engineer often decided he or she could help after all. Her purpose was not to accuse anyone of bluffing, but simply to make sure that requests were fully understood.

In this effort, Bob Mitchell, her boss and the current JPL *Cassini* program manager, was a stalwart. Bob took over from Spe in 1999 to lead the Operations phase of the project at JPL. He helped lay down the science sequences in which the different instruments would observe Saturn's system, but built in a "surprise factor" in case something was so unexpected and spectacular that the scientists wanted to take a second and third look. In doing so he placed himself squarely and skillfully between the columns in the box about engineers and scientists, planning yet exploring, following procedures yet being prepared to alter these if necessary.

> "We will work with the scientists to get there," Bob would say, and that was a real change as too often we had heard from some of the engineers responsible for operating the space craft, "That is not in our requirements."
>
> Candy Hansen, JPL investigation scientist

Thanks to the very special skills of those involved, the *differences* between engineers and scientists became a *dialogue*, wherein each came to appreciate the other's contribution.

Europe as a partner with additional resources

The relative importance of scientists and engineers has everything to do with the level of participation in the joint mission by European experts. Half the scientists are Europeans, while a majority of the engineers directly on the project are Americans, with the notable exception of the *Huygens* engineering team and ESA's Darmstadt Operations engineers.[5] Moreover, Europeans had not grown up in a world were the concern was winning a space race; they rather sought to unify their Community through shared knowledge.

Contrasting forms of funding in the United States and Europe had the effect of favoring European science, as we saw in Chapter 1. While American scientists are funded directly via Congress and NASA, European scientists are funded by the nation from which they come, and they brought their funds with them. VIMS scientist Kevin Baines not only appreciated that Europeans were "free," he felt that "it was a great boon to have them. They put out great work." Dennis Matson was in no doubt, "the funds shortage helped to spread the work to Europe." This was possibly to the chagrin of many US scientists who did not get a seat on this bus as a result.

Wes Huntress believes we should approach the solar system "as a planet." The United States has to put trust in a partner, who is part of the critical path to a joint objective:

> That is what I liked about this mission. But I think NASA does not like this any more. [It's] too hard, too complex, and [there's] too much reliance on persons not under its direct control. Yet partnership leads to better science and engineering. All those different approaches, cultures, and sources of knowledge tend to lead to better results.

One obvious advantage on the side of both Europe and science was that the highlight of the whole mission, and its most far-reaching and ambitious aspect, was the landing of the *Huygens* probe through the orange mists of Titan. The probe was Europe's special contribution, and a high-risk undertaking, with parachutes packed tightly for almost ten years opening in an atmosphere very different from that of Earth at temperatures far below freezing. ESA's engineers had never before designed a probe, and – as we have seen – European planetary science had led the quest. While the orbiter continues its trajectories after the probe's release and descent into Titan's arms, this phase was arguably the climax of the drama, the penetration of the veiled "planet" long hidden from our eyes. Space science has rarely encountered so auspicious an arena for discovery. As the time of landing approached, devotees the world over held their breath in anticipation.

The crisis of restructuring: losing the platforms

With the 1991 cancellation of CRAF, *Cassini-Huygens*' sister program, and the pressures on NASA's budget, the mission found itself extremely vulnerable for a while. It had only recently been approved, but the US government was looking for cuts, and it made an inviting target for the cost-cutters.

It fell to John Casani, Dick Spehalski, and Tom Gavin to do emergency surgery on the spreading bulk of the whole project. The stratagem of slowing down so as to spend less per annum would no longer be tolerated. The whole mission had to cost less, or NASA would cease to support it. The space scientists in Europe and America were very, very worried, and for good reason. Their dreams hung by a thread. "That restructuring was the hardest thing I've ever done," Dick Spehalski told us. Anyone looking for an issue that would set engineers and scientists at each other's throats needed to look no further. Where would the cutbacks be? Dick was convinced that cutting back on the mission's technical capability was not possible, or the spacecraft would simply fail to get to Saturn and all the science would be lost.

But if the capabilities for reaching Saturn were already stretched, what else would have to give? Attention now centered on the movable platforms, as we have seen, and getting rid of them would certainly save money, at least during those years when the *Cassini-Huygens* project was most threatened with cancellation. This would certainly help prevent the loss of the entire project, but it would also carry a cost for scientists in preparation for and during the observation period at Saturn.

It had only been a few years since the announcements of opportunity had been issued by various agencies, and the principal investigators had celebrated their successes. Now the opportunity to make observations was being cut back, but the only realistic alternative was to cut out one or more of the instruments. It seemed to both the scientists and the engineering team that this would be a worse alternative, in the context of a once-in-a-lifetime mission of this magnitude.

> The credit for realizing what had to be done to get rid of cost, and to keep as much science as possible, goes to John, Spe, and Tom. And to their credit as well, they involved the science team in this decision. It was a decision nobody liked, which is probably why it was successful.
>
> *Wes Huntress, former NASA associate administrator of space science*

What upset the scientists most was that it was doubtful whether any real savings would be made by abandoning the platforms. It would help cut the budget requirements in the early 1990s, but it would mean that it cost more to operate the whole craft from 2004 onwards, and much time would be wasted in orbit by slow reorientations of the craft so that the instruments could point in the right direction to suit different observations and their teams. As we have seen, many of the scientists we interviewed still could not forgive this decision, which forced the teams around each instrument into lengthy negotiations about sequences of observation.

Other cost savings came from other actions the leaders took to create a far more efficient project, in an attempt to minimize the impact on the scientists. Together, these actions saved the program.

However, some felt that more could have been done to compensate for these design changes. Bob Mitchell described the spacecraft as having some technology that was state-of-the-art at the time it was developed, but the lack of movable platforms to point instruments even predates *Voyager* capabilities. " We did lose some science capability as a result and that makes it harder to now meet everyone's expectations."

Chris Jones, now director of planetary projects at JPL, was a key player in the decision to get rid of the platforms, and go to the "*Magellan* mode." He described it as "a huge sacrifice to swallow for the scientists, but the *Magellan* mission had used this design successfully and we were still going to get ten times as much data as *Voyager* had done."

E-mails were angrily exchanged between some scientists and sent to NASA about JPL engineers helping themselves to diminishing funds in their own interests; were the engineers trying to sideline the interests of the scientists? Were they trying to wrest the initiative, while the scientists were helpless spectators waiting in the wings? Why was a big project conference called off after several European colleagues were already airborne over the Atlantic? To be precise about all this, the changes to the craft and the project overall were not really an international issue. They mostly affected the European scientists on several of the *Cassini* instrument teams; *Huygens* and its development and science were largely unaffected.

With the hard decisions made during the de-scoping phase in 1991/2 behind them, Dick Spehalski and Dennis Matson moved quickly to reassure the scientists that their interests were not being ignored. For example, they instigated a process by which all engineering and science instrument teams had to take full responsibility for their budget allocation (with no further future reserves available at all). Dick obligated his engineers to pay scientists for the ramifications of any unplanned and late changes. Scientists had been promised a certain budget, a certain amount of power for their instruments, and a certain amount of room or "mass" aboard the orbiter.[6] (See also page 96.) Any change of mind by engineers, even if for good reason and unforeseeable circumstances, had to leave scientists as well off as they had been before. Hence any encroachment by engineers on the scope to carry out the science, occasioned by the restructuring, now had to stop. Alterations to instruments or observations had to be paid for by engineers from their own budget, so that

> [Not all scientists were all that unhappy about the platform decision.] For the imaging observations of Titan we need very long exposure times; we need to collect photons for a very long time, and for that the instrument needs to be very steady. By being bolted to the spacecraft, you are inherently very steady, probably more so than otherwise would have been the case. So for imaging Titan's surface, we are actually a lot better off this way.
>
> Alfred McEwen, University of Arizona, member of the ISS instrument team

they bore both sets of costs. Moreover, the mass and power available to scientists could not be further curtailed.

Engineers' requests for changes came to Tom Gavin, head of the Spacecraft System Office, and he would go over these with the scientists, first convincing himself of their necessity, then helping to convince others, yet making sure that science would not lose. For example, if an error by engineering in vibration specs could only be fixed by weight and cost increases, the engineer had to restore these resources to the team conducting the observations. Spehalski was urged to add more nuclear power to the craft, but made the difficult decision to encroach no more upon the existing agreements and arrangements.

The cost-cutting exercise and platform elimination were not without their advantages, some of them unexpected, as we have seen. Principal investigators and facility instrument team leaders (the difference between these is explained in note 13 on page 201) were forced into prolonged negotiation with the Development and Mission Operations engineers, which is still going on today for the remaining years around Saturn. From this they learned much more about the interdisciplinary nature of the whole mission.

The effect was to bring to bear the influence of scientists at a much earlier juncture, since they could advise on how engineering designs would affect research. Instead of engineers and scientists being joined at the apex, all now had to work together on the smallest minutiae, and thus they became increasingly well informed on what the craft was supposed to accomplish. Most especially the JPL engineers now felt impelled to serve scientific objectives, and set out to convince their colleagues of their dedication to the larger cause.

This is not to say that this project's scientists and engineers have always been very happy with each other. In the last years before launch, the emphasis was appropriately on getting the craft and rocket safely off the ground and on the way to Saturn. We also heard scientists say that the change in project leadership after launch was well received in terms of transitioning the emphasis at that juncture to the science objectives of the mission – aligning leadership roles with the changing needs of the project.

What was unusual about this mission – particularly compared with the previous similarly large *Galileo* mission to Jupiter – was how early the science

> This has been a hard pill to swallow for these guys. I decided to be "open book" and had the engineers explain in two days to the scientists what they were doing here in terms of space craft programming, navigation, sequence development, and uplinking. I wanted them to know that we were not wasting money here that should have gone to the scientists. I gave all of them schedules and milestones so they had complete insight into the engineering activities that were being done in support of acquiring their science data. Sharing this information went a long way toward solving the problem, and the relationship between the scientists and the engineers has been one of constructive teamwork since.
>
> *Bob Mitchell, current JPL* Cassini *program manager*

teams and spacecraft development engineers were brought together to influence each other. Dennis Matson believes that much grief can be, and was, avoided by articulating the science objectives of the mission to Saturn very early on, so that science could be a crucial part of mission design from the very outset. In many ways, scientists and engineers co-designed the hardware and the capabilities it needed.

Charley Kohlhase, *Cassini* science and mission design manager, pointed out that JPL's *Cassini-Huygens* mission design approach had always been a cross-weave of science and engineering: "It was all about knowledge return, and always had been."

The US team approach also influenced the *Huygens* development team at ESTEC. This was exemplified by Bernardo Patti, *Huygens* thermal and mechanisms engineering manager for ESTEC, who told us in his charmingly modest and ironic way about his own role: "I see myself as a very sophisticated bus engineer. What makes it worthwhile is to know where the bus is going and why."

Setting up the trading system

One issue between engineers and scientists nearly wrecked relationships between them on *Galileo*, and also haunted the doomed *Mars Observer*. When scientists respond to announcements of opportunity, they must estimate their observational requirements, the costs of their instruments, the power they will use, the mass or space they will occupy, and the data rate of transmission needed. They are likely to win a place aboard the orbiter if their proposed investigation is scientifically judged to be profound, and also if their demands for resources are sufficiently modest in scale given the total resources available.

In several previous missions, the proposing scientists generally underestimated the resources needed. Once their instruments had been selected and they were nominally "on board," the scientists would announce that more resources had turned out to be necessary. (Sometimes they initially overestimated their resource requirements, but this was less likely since they were more liable to be selected if their requirements seemed modest. When this happened they tended to stay silent. After all, the engineers might later try to reduce their resources, so it made sense to have a margin of safety.)

Anticipating these clamorous demands and shortages, the Spacecraft Systems Office kept a reserve of resources, but the principal investigators and team leaders realized this and petitioned early and often to get more resources allocated to them. On the earlier missions this pitted scientists *against* each other, weakening them further in comparison with the engineers, who were among those who arbitrated between their demands. These conflicts and shortages became so serious that a near-infrared mapping spectrometer (NIMS) was forced off the *Mars Observer* entirely, leaving no instrument capable of detecting water on Mars![7]

This particular type of problem has a long history, and is sometimes called after the medieval Tragedy of the Commons. In medieval times European countries generally had stretches of common land (that is, land owned by no one) upon which peasants could graze their animals. For many years this system worked well, especially after the Black Death, which cut the European population by a third and so reduced the demand for land. But as the population increased, pressure on land resources grew intense. Those with enough power to do so began to enclose land for their exclusive use (effectively claiming ownership of it, and removing it from the common stock). This in turn accelerated the chronic shortage of common land.

The point is that many "free" resources tend to become over-utilized. When the utilization reaches a certain level, the system can switch suddenly and catastrophically from serving everyone's interests to serving few if any interests.

The same thing happened in JPL's Spacecraft Systems Office. Much of what was actually spare resource capacity was allocated in the way described above, and became hidden and so unavailable, shutting out those in genuine last-minute need.

John Casani had learned of a trading system (see the box, "The instrument with the black hand") that would allow principal investigators to trade resources they held in surplus against resources in short supply. The Spacecraft Systems Office would no longer hold any reserves for instrument scope changes, so when the instrument teams had over- or underestimated their needs they would have to trade with each other for what they needed. There were additional reserves, but only for changes induced by the spacecraft or project itself.

> If we institute this trading system, we should rename this mission as *Casino-Huygens*.
>
> Jean-Pierre Lebreton

Initially, the system introduced by John, with Spehalski's and Dennis Matson's strong support – it was another element in their drive to keep costs down and safeguard the mission – had to fight against deep suspicion. Market trading systems are associated with right-wing economics, and many scientists regard economics as a pseudo-science at best, replete with ideological excitements.

Yet on closer examination none of these fears were born out. This was technically an "after-market" – resources had been allocated originally on the basis of scientific merit and need, not on the capacity to trade (although some did just that). The trading had to do with later adjustments to those allocations, occasioned by uncertainties about the size, cost, and power needed by each instrument. There were almost bound to be imbalances between estimates and eventualities, and it was these that the instrument teams traded among each other. The only real question, given these differences, was whether they should be adjudicated or traded? (See the box, "Adjudication versus trading.")

The instrument with the black hand

The term "black hand" – or *mano nero* – originated from the extortion rackets run by immigrant Sicilian and Italian gangsters in the Italian communities of New York City, Chicago, New Orleans, Kansas City, and other US cities from about 1890 to 1920. It referred to gangs sending threatening notes to local merchants and other well-to-do persons, printed with black hands, daggers, or other menacing symbols, and demanding money on pain of death or destruction of property. The black hand declined with the entry of Prohibition and big-moneyed bootlegging. Bootlegging and space science are far apart, but the VIMS instrument team was at some point figuratively marked with the black hand.

In the late 1980s John Casani was responsible for the *Mars Observer* program, a mission that came to an unfortunate end in 1992. That project suffered from major cost overrun problems in instrument development and science planning, which eventually led to a decision to eliminate a VIMS instrument from the *Observer* spacecraft altogether and to de-scope several other instruments. John knew that something had gone wrong in the *Mars Observer* project, but was not sure exactly what in JPL's project management approach had caused these overruns. Fearing the same would occur with the *Cassini* project, he asked internal JPL staff to analyze recent space missions. This led to a study commissioned to the Division of Humanities and Social Sciences at the California Institute of Technology, conducted by Professor John Ledyard, assisted by Professor David Porter and doctoral student Charles Polk. Their findings raised the "Commons" phenomenon and the "insurance" issue described in this chapter. John asked Professor Ledyard and his colleagues some very hard questions, and as a result they jointly created the instrument exchange system formally called the *Cassini* science management plan. John knew that JPL could hardly afford similar overruns to those on *Mars Observer*: the costs of the initial CRAF/*Cassini* program were already mounting, and NASA headquarters was sending clear signals that the Cassini *team must* reduce costs and keep to a fixed budget.

When Bob Brown was selected as the facility instrument team leader for VIMS, it was stated in his selection papers that if the project ran out of money he and his instrument would be dropped. Initially he hesitated to take the assignment, knowing what had happened with *Mars Observer*. John convinced him and the team scientists that this time things would be different, and they were. The team's Science Instrument Office leader Bill Fawcett, along with Spe and Dennis, "managed the process beautifully," remembers David Porter.

Ironically, it was the exchange methodology itself that received the black hand years later, in 2003, when Admiral Poindexter tried to predict the probability of terrorism at the Pentagon with a proposed total awareness information trading system. He lost his job over it and embarrassed his boss, Donald Rumsfeld. Congress and the press played the role of the *mano nero* messengers.

Adjudication versus trading

Allocation with adjudication		Allocation with trading
Power at the center	◐	Power decentralized
Hierarchical	◐	Egalitarian
Decisions are political/administrative	◐	Decisions are science based
Rivalry for scarce resources	◐	Exchange of scarce resources
Winners and losers	◐	Win–win solutions
Substantial sacrifices	◐	Scant sacrifices
Three-cornered relationships	◐	Bilateral relationships
Truth is difficult to gauge	◐	Truth reflected in bidding
Take and take	◐	Give and take
Trade-offs are hard to estimate	◐	Price reflects genuine mutuality
Resource hoarding	◐	Resource sharing

In the first scenario power is centralized, with engineers among the chief decision makers, enabling them should they wish to *divide* and *rule* the scientists. The second scenario, trading, is far more egalitarian, and allows decisions to be made based on science, not administrative whims. Moreover, the resources are now less scarce, because those who once held a surplus over requirements in one resource can now trade these for what they really need. The number of win–win solutions rises, and the need for sacrifice falls. Bilateral relationships between scientists help maintain their friendships as well as their cohesion and capacity to champion science. Instead of bluffing about the resources individuals "must have," truth is reflected in the price they are willing to pay. They no longer hoard, they share.

One positive result of this trading system was that none of the twelve instruments on *Cassini* had to be canceled. All assumed individual responsibility for mobilizing sufficient resources to make their instrument viable. No one tried to blackmail the Spacecraft Systems Office. David Porter and Randii Wessen, who helped design and operate the trading system, explain:

> The *Cassini* Resource Exchange is a computerized, multi-dimensional barter system that resides on the Internet. As a point of history, it was in fact the first "real" trading system that used the Internet as a communication network.... During the early history of the trading system (late 1993 to early 1995) bidding and trading were brisk. Results from the system show 29 successful trades, all but two involving Budget and Mass. Those involving current fiscal year (FY) funds for future FY funds, we call money market trades.... In the mass market there were eleven contracts with over twelve kilograms exchanged.... Very few bids were tendered for Power and only two trades occurred.

That this system was initiated by top engineers of the likes of John Casani and Dick Spehalski, and that its chief beneficiaries were scientists, goes a long way to explaining the much-improved relationship between the two disciplines in this project. It was a win–win situation, in that the entire project and its management benefited. It worked so well that only once did Dick have to make an exception – when a European scientist did not deliver as promised and the instrument team had trouble as a result.

Science gains over time

We cannot really appreciate the integration of engineering and science achieved by *Cassini-Huygens* unless we also look at this phenomenon over the length of the mission. The first practical question has to be, how do you get there and at what cost? This meant that engineers had to prepare for the launch, engineers had to help design the scientific instruments, and engineers had to help plot the circuitous route to Saturn. Some suggested that it was the marked improvement in rocket technology in the early 1990s that permitted so much science to be taken on the craft. Had this not occurred, all attention might have been focused on "making it" to Saturn (see Chapter 6).

As it was, the launch date was postponed more than once, and getting the mission launched in the brief window of opportunity available preoccupied nearly everyone. Subsequently the Earth and Venus fly-bys were the focus of attention. Since there would be a seven-year interval between launch and arrival at Saturn, it was decided to do the science sequencing after the launch, not before it.

Indeed, if we look at the sheer length of time taken building the craft and flying it, this comes to some fourteen years, during which time nearly all the science remained to be done. Candy Hansen remembers that:

> For a lot of those years no one wanted to hear about science operations, while admitting they were crucial. It was not until we were two years away from Saturn that people began to wake up to what was surely the most crucial phase of all. Partly the delay was because the waiting scientists were "distributed" across the globe, getting on with their careers.

The mission staff also became preoccupied at this stage with developing distributed information systems for the collection and dissemination of the science data to come. JPL for the first time became the coordinating rather than the controlling institution – a sign of how much *Cassini-Huygens* had become a worldwide cooperative science expedition.

Following the launch, a majority of the engineers went on to other things, and the relative numbers of engineers and scientists began to shift in the latter's favor. From the successful launch onward, it gradually dawned on people that the science itself at Saturn and Titan was the final measure of success, and this still lay ahead. Science was to come into its own at last.

Nothing could have been more fortuitous than the fact that the team had a chance for a dress rehearsal at Jupiter. The reason for flying by Earth, Venus, and Jupiter was not to study these but to borrow their momentum and be flung onwards into space. Yet as the spacecraft neared Jupiter with most of its crises behind it, it occurred to several people that not only could important images of the planet be captured, but here was a valuable opportunity to rehearse a partnership. The engineers of Mission Operations and the scientists around each instrument had to get their acts together if the promised data were to be captured.

Candy Hansen and her colleague Scott Bolton, JPL investigations scientists and co-leaders of the TOST team, were major advocates for making the Jupiter fly-by a test ground for later cooperation around Saturn. Now they could sort out what the *real* constraints were, as opposed to just putting blame on excuses invented by engineers. It was not an easy task. Images of Jupiter were "not official requirements," the spacecraft team informed her. But she persisted:

> We had to be careful what we asked them to do, because they could have said "no" at any time and then made a habit of it. But it was a cold water shock to have to actually develop a command sequence in which instruments were pointed and had to take their turns.

It was a relief to stop talking and arguing and actually *practice* sequencing. It was not easy, but spectacularly successful once they got into the hang of it. The craft had a sequence of set maneuvers, which would give all instruments a chance, and these were repeated several times. Investigators and instruments fitted themselves into the pre-defined slots. What raised morale sky-high was that no one had counted on this "rehearsal" becoming an additional learning opportunity.

After years of expectations being damped down by engineering priorities, here was a science bonanza that no one had expected – and the spaceship was still two years from its rendezvous with Saturn. But what the scientists appreciated most was the way engineers in Mission Operations now threw themselves into the challenge, after having been much less flexible during the development years. Candy testified to the sudden change of attitude by scientists towards engineers. "Wow! These guys really work hard to make our investigations better. We really appreciate what they do!" These engineers directly facilitated the

> We had a fabulous fly-by with great data. People were so happy with what they had, that no one thought about what they did not get.
>
> Scott Bolton and Candy Hansen, JPL investigations scientists and co-leaders of the TOST team

scientific observations. They now realized, if they did not know already, that discovery was the purpose of the entire enterprise.

The closer the craft approached to Saturn, the more all concerned identified with science exploration. Kathryn Weld, manager of JPL's *Cassini* Science and Uplink Office, began to notice things that must have been latent there all along:

> Some of the engineers have built their lives around Mission Operations. They sometimes stay all night, often sleeping on the floor so as to miss nothing. You would think they were scientists themselves, so much did they care about knowing. They won't even get the credit for these observations, the scientists will, yet I never felt less important than the scientists and was never treated so. This was what years of development work had all been for, and we were all eager to witness the culmination.

Chris Jones, currently director of planetary projects at JPL, felt that as the mission reached its climax, engineers whose concerns had dominated for so long to get the craft to operate successfully yielded gracefully to the science at the core of the whole project:

> Engineers at JPL are not about engineering per se, but about improving performance to achieve science objectives or squeeze more science out of what we can bring. The same was true of the development phase. It was always about creating a stable system which could do science. Science textbooks will show pictures of the planets, not the engineering feats needed to get them there, so in practice we achieved a balance between the needs of both.

Although much may have been lost by compromises along the way, dramatic advances in communications made up for this. While the admittedly handicapped *Galileo* had transmitted at 8.33 bits per second, *Cassini-Huygens* was transmitting at thousands of bits a second. Those observations lasting only a few hours might not be so barren after all. Science had come into its own, with engineers working night and day to cheer it on.

Qualities emerging from paradox

While planetary science had long been the official reason for space programs, in practice many other agendas elbowed it out of the way. So long as the reliability of spacecraft was the chief worry, science had to often take a back seat.

But somewhere along the way in this *Cassini-Huygens* mission, science caught up with engineering. This is one more reason that our experiences up there at Saturn, although millions of miles away, can tell us something about ourselves. *Cassini-Huygens* defied the odds to reconcile the *linearity* of engineering with the *circularity* of science, coming full circle to engage the unknown. All were stimulated to carry on a dialogue between the disciplines, with appreciative understanding of their differences and sometimes grievances. Together they overcame the stresses, with the JPL engineers in particular going out of their way to restore confidence among the scientists at several stages of the project.

In a very real sense, science completes engineering. The former is the latter's destiny: not fighting nature but engaging it. Engineers must *master "getting" there* so scientists can *unleash their curiosity*. Reaching Saturn is both an end in itself and a means to scientific ends. Through careful conservation of hardware you can better exploit its capability. Having been *pragmatic*, the team is now free to *speculate*. Engineers design and develop a spacecraft for artists and explorers to utilize. They operate in a predictable manner so that surprises may be sprung later.

Taking *linearity–circularity,* and *predicted results–boundless speculation* as typifying the many paradoxes we have seen in this chapter, the learning loop reads like the one in Figure 5.1. It does not take rocket science to see this too is a developmental helix towards "engineered discovery," a reconciliation of these paradoxes as illustrated in Figure 5.2 This dynamic is broadly applicable to interfaces between the two groups. These inherent strengths must have contributed to the project's ability to overcome the crises it faced, as we will see in the next chapter.

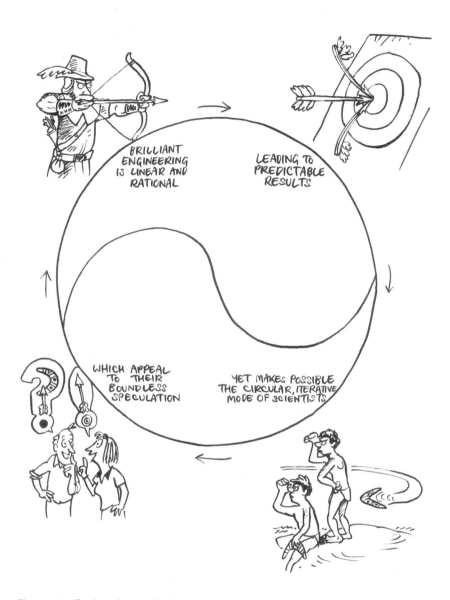

Figure 5.1 Engineering and science

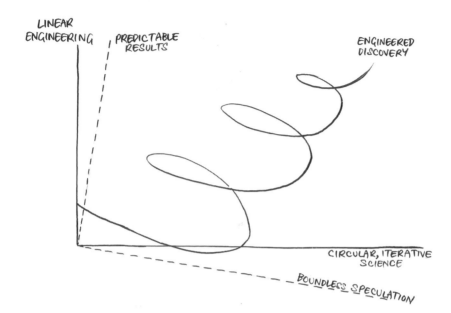

Figure 5.2 Engineered discovery

Taking a closer look

It is October 15, 1997, and *Cassini-Huygens* starts its seven-year, more than 2 billion mile (3.2 billion kilometer) journey to Saturn. The craft's full weight is over 12,000 pounds (5,400 kilograms), so large and heavy that there is no available launch vehicle that could send it on a direct path to Saturn, so it will be a long and winding road. The spacecraft will use an interplanetary trajectory that takes it by Venus twice, then past Earth and Jupiter. The planets tugging on the craft will help hurl it towards Saturn.

On April 26, 1998 *Cassini* swings by its first port of call, Venus, which speeds it up, aided by firing its engines – what is known as a deep-space maneuver. Then 423 days later it is again very close to Venus, only some 370 miles (590 kilometers) away, for an additional kick, all the while protected by the high-gain antenna against the scorching heat of the Sun.

Cassini now turns away from the Sun on the long track to Saturn. On August 18, 1999 the spacecraft pays Earth a final, brief visit, swinging past the mother planet at an altitude of only 727 miles (1,163 kilometers), close indeed but carefully kept away from any collision. It now has enough momentum to reach the outer solar system. Passing Mars from a distance and then through the asteroid belt, *Cassini* approaches the giant Jupiter for a final gravity assist on December 30, 2000. Although *Cassini* stays some 6 million miles (9.6 million kilometers) away from Jupiter's cloud tops, the planet's pull is strong, and gives the craft one more boost and a small change of direction for its final track to Saturn, the sixth planet from our Sun.

The Jupiter fly-by poses a good opportunity for *Cassini* and the already present *Galileo* spacecraft to jointly study several aspects of Jupiter and its surrounding environment from October 2000 through March 2001. The scientific observations take advantage of having two different vantage points in the vicinity of the planet at the same time. Working in tandem, the magnetospheric imaging instruments (MIMI) on both *Cassini* and *Galileo* map Jupiter's enormous magnetic field in three dimensions, as high-energy particles from the solar wind shape its boundaries. We learn more about the auroras at Jupiter, while the "eyes" of the imaging sub-system (ISS) send a wealth of dazzling pictures back to Earth. At Jupiter alone, ISS has acquired about 26,000 images, among which are previously unseen emissions from the moons Io and Europa, and a volcanic plume over Io's south pole.

The studies serve as good practice for checking out many of *Cassini*'s instruments, and most of the *Cassini* orbiter science instruments are turned on and calibrated during this encounter, preparing for the real thing later, while scooping up as much scientific gravy as possible in this neighborhood. The giant disk-shaped radio antennae in the Mohave Desert and in Canberra and Madrid – Earth's Deep Space Network – take in all this wealth of data, while facilitating a constant dialogue with the craft and instruments.

The ship travels for another three and a half years or so before arriving at its destination, Saturn, from where radio signals take a round trip to Earth of over two and a half hours – in lay terms, we do not know what happens until more than 75 minutes after it actually took place. *Cassini* is now plunging towards Saturn at a speed approaching 50,000 miles (80,000 kilometers) per hour, passing the moon Phoebe on its way several weeks before arrival.

The most critical phase of the mission after launch is Saturn orbit insertion on July 1, 2004 (universal time; it occurs on June 30 in US time zones). The spacecraft fires its main engine for 96 minutes starting at 7.36 pm Pacific daylight time, to break its enormous speed to a still very respectable 1,300 miles (2,100 kilometers) per hour, thus allowing it to start its orbit around Saturn, first flying above the plane of the magnificent rings, then passing through the gap between the F and G rings. The high-gain antenna is positioned to act as a shield to protect the craft against small particles that may be present in the region of the ring crossing. It all seems like a movie, but it is very real, and the precision required is mind-boggling: one degree off, or too much speed, and *Cassini* would be forever lost into deep space.

But the craft arrives safely, and begins the first of the 76 orbits to be completed during its four-year primary mission – the primary target, which could be exceeded, as there may be fuel left for two more years of this extravaganza. An extravaganza it will be, as the science payload is expected to haul in several trillion bits of data.

The arrival period provides a unique opportunity to observe Saturn's rings and the planet itself, as this is the closest approach the spacecraft will make to Saturn during the entire mission, at about 12,000 miles (19,200 kilometers) above the planet's cloud tops. Understanding the rings is one of the quests, as their origin remains uncertain. Are they left-over material from the formation of our solar system that never coalesced into a larger body, or remains of moons pulverized by meteors? What creates their color if they are mainly made out of water ice? What causes the mysterious spoke patterns to flicker on and off?

Being so close allows the dual-technique magnetometer (MAG) – mounted on a boom to isolate it from the craft's interference – to measure the strength and direction of the magnetic field, which may give us clues about the structure of Saturn's core and interior. MAG has been deployed since the craft got under way, and measured the magnetic fields of the planets the craft passed on its tour. The radio and plasma wave spectrometer (RPWS) listens for evidence of lightning above the planet, and searches for large meteoroid impacts on the rings. Several of the other instruments are involved to scrutinize the rings in detail, such as the cosmic dust analyzer (CDA), which is designed to help scientists discover whether the cosmic dust in the vicinity of Saturn (this dust is roughly the size of cigar smoke particles) stems from the rings. The pictures sent back to Earth are incredible both in what can be seen and in their clarity.

For the next six months, *Cassini* completes three perfectly executed orbits of Saturn. Instruments are now taking turns to do their science observations. The magnetospheric and plasma science instruments begin to monitor the solar wind; the ultraviolet imaging spectrograph (UVIS) observes Saturn's aurora and inner magnetosphere, and the imaging subsystem (ISS) takes a look at the southern hemisphere of the moon Iapetus. The optical remote sensing instruments examine Saturn's south pole and aurora, while the visible and infrared mapping spectrometer (VIMS) and the ISS's cameras take mosaics and movies of the rings and Saturn's south pole. VIMS' infrared (literally: "beyond the red") radiation reads the fingerprints of each chemical component it observes, and helps identify what gases and surface materials are being observed.

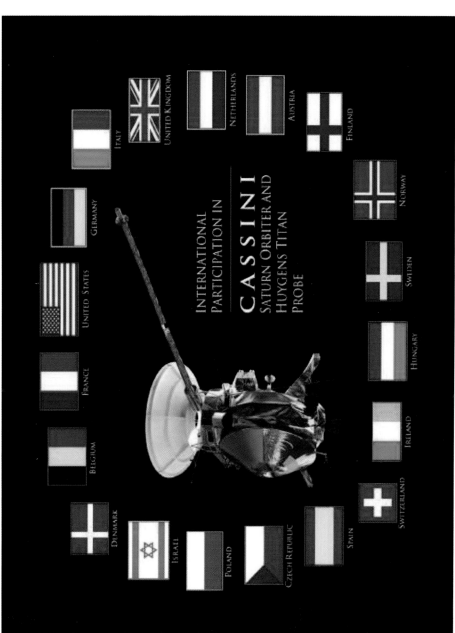

Image 1 International cooperation: these flags illustrate the international scope of the *Cassini-Huygens* program. Although its primary sponsors are the National Aeronautics and Space Administration (NASA), the European Space Agency (ESA), and the Italian Space Agency (ASI), the *Cassini-Huygens* team encompasses academic and industrial partners in 33 US states and 19 nations.

Courtesy NASA/JPL-Caltech.

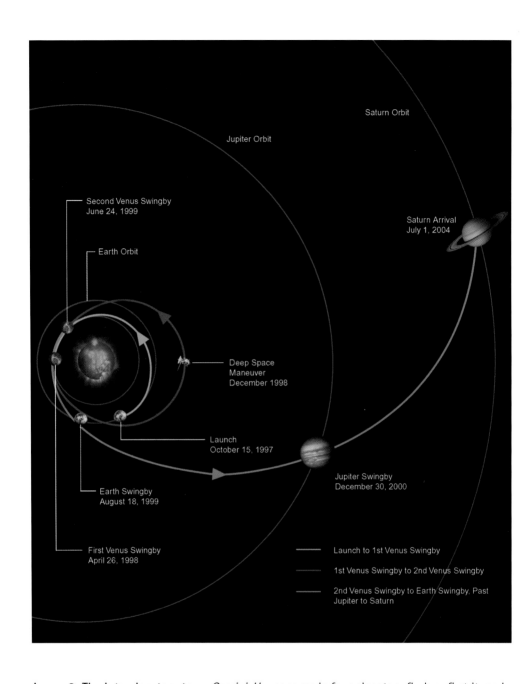

Image 2 The interplanetary tour: *Cassini-Huygens* made four planetary fly-bys: first it made two fly-bys of Venus (April 28, 1998 and June 24, 1999) and, next, one of Earth (August 18, 1999). These gave the spacecraft enough energy to send it on a path towards Jupiter, where it received its final boost to Saturn. The fly-by of Jupiter occurred on December 30, 2000. *Cassini-Huygens* arrived in orbit around Saturn on July 1, 2004.

Courtesy NASA/JPL-Caltech.

Image 3 **Assembling the *Cassini* at Kennedy Space Center:** JPL technicians reposition and level the *Cassini* orbiter in the Payload Hazardous Servicing Facility at the Kennedy Space Center in July 1997, after stacking the craft's upper equipment module on the propulsion module.

Courtesy NASA/JPL-Caltech.

Image 4 **Attaching *Huygens* to *Cassini*:** The *Huygens* probe is installed into the *Cassini* orbiter in the Payload Hazardous Servicing Facility at the Kennedy Space Center in July 1997.

Courtesy NASA/JPL-Caltech.

Image 5 Lift off! The *Cassini* spacecraft and *Huygens* probe begin their seven-year journey to the ringed planet. The successful launch aboard a Titan IVB/Centaur occurred at 4:43 a.m. EDT, October 15, 1997.

Courtesy NASA/JPL–Caltech.

Image 6 **Leaving Earth:** this picture of the *Cassini* launch was taken by Ken Sturgill of Marion, Virginia, using a 30 second, f 1.8, exposure on 400 speed film.

Courtesy NASA/JPL–Caltech and Ken Sturgill.

Image 7 **Jupiter fly-by:** The solar system's largest planet, Jupiter, and its moon, Ganymede, in a color picture taken by *Cassini* on December 3, 2000. Ganymede is larger than the planets Mercury and Pluto and Saturn's largest moon, Titan. *Cassini* was 16.5 million miles (26.5 million kilometers) from Ganymede when this image was taken.

Courtesy NASA/JPL/University of Arizona.

Image 8 **Phoebe close-up:** First images from the *Cassini-Huygens* fly-by of Phoebe, another of what may turn out to be no less than 46 moons of Saturn, based on recent discoveries, reveal it to be a scarred, cratered outpost with a very old surface and a mysterious past, and a great deal of variation in surface brightness. There is a suspicion that Phoebe, the largest of Saturn's outer moons, might be parent to the other, much smaller retrograde outer moons that orbit Saturn. Six high-resolution images taken of Phoebe by *Cassini* have been put together to create this mosaic.

Courtesy ESA.

Image 9 **The magnificent rings:** *Cassini-Huygens* pierced the ring plane and rounded Saturn on October 27, 2004, capturing this view of the dark portion of the rings. A portion of the planet's atmosphere is visible here, as is its shadow on the surface of the rings. The image was taken with the spacecraft's wide angle camera at a distance of about 384,000 miles (618,000 kilometers) from Saturn.

Courtesy NASA/JPL-Caltech.

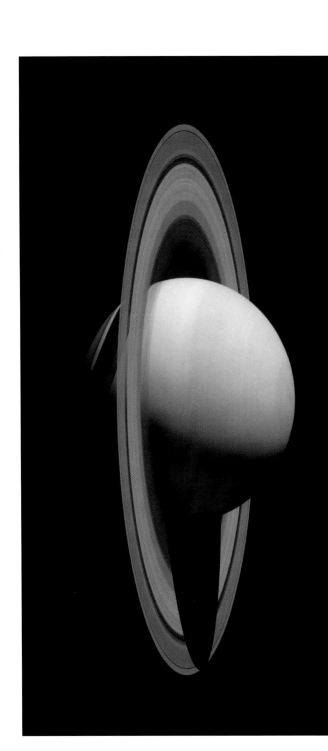

Image 10 **Saturn's beauty:** While cruising around Saturn in early October 2004, *Cassini* captured a series of images that have been composed into the largest, most detailed, global natural color view of Saturn and its rings ever made. This grand mosaic consists of 126 images acquired in a tile-like fashion, covering one end of Saturn's rings to the other and the entire planet in between. The smallest features seen are 24 miles (38 kilometers) across. Among the features seen are subtle color variations across the rings, the thread-like F ring, ring shadows cast against the blue northern hemisphere, the planet's shadow making its way across the rings to the left, and blue-grey storms in Saturn's southern hemisphere to the right. Tiny Mimas and even smaller Janus are both faintly visible at the lower left. The Sun–Saturn–*Cassini*, or phase, angle at the time was 72 degrees; hence, the partial illumination of Saturn in this portrait.

Courtesy NASA/JPL/Space Science Institute

Image 11 **The rings' kaleidoscope:** Images taken by *Cassini-Huygens* on June 30, 2004 show the compositional variation within the rings. From bottom to top, this picture shows the outer portion of the C ring and inner portion of the B ring, which begins a little more than halfway up the image. The general pattern is from "dirty" particles indicated by red to cleaner ice particles shown in turquoise in the outer parts of the rings. The ring system begins from the inside out with the D, C, B and A rings followed by the F, G and E rings. This image was taken with the ultraviolet imaging spectrograph instrument, which is capable of showing features up to 60 miles (97 kilometers) across.

Courtesy NASA/JPL-Caltech.

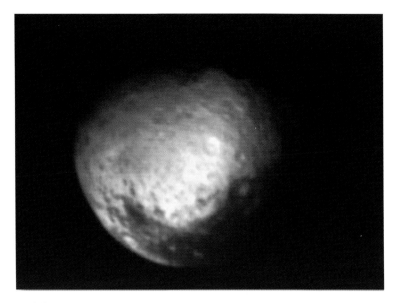

Two moons of Saturn: Image 12 above, Iapetus and **Image 13** below, Rhea.

The leading hemisphere of Iapetus (the side that faces the direction of rotation) is very dark, reflecting only about 4 percent of the sunlight that falls on it. The trailing hemisphere reflects about 50 percent of the sunlight that falls on it.

At 950 miles (1,528 kilometers) in diameter, Rhea is the second largest satellite of Saturn. Only Titan is larger. Bright wispy streaks cover one hemisphere of Rhea.

Both courtesy NASA/JPL-Caltech.

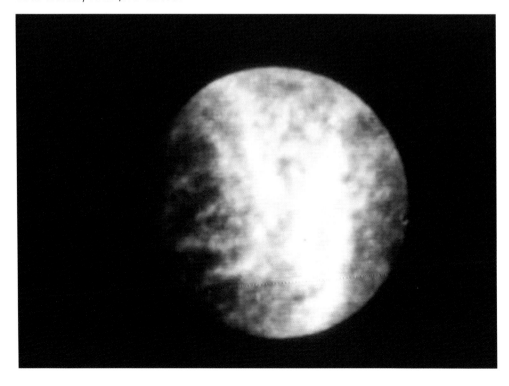

Image 14 **Titan's atmosphere:** this false-color composite of Titan was created with images taken during *Cassini*'s closest fly-by on April 16, 2005. It was created by combining two infrared images with a visible light image. Green represents areas where *Cassini* is able to see down to the surface. Red represents areas high in Titan's stratosphere where atmospheric methane is absorbing sunlight. Blue along the moon's outer edge represents visible violet wavelengths at which the upper atmosphere and detached hazes are better seen. North on Titan is up and tilted 30 degrees to the right. The images used to create the composite were taken with the *Cassini* spacecraft's wide angle camera, at distances ranging from approximately 107,500 to 104,500 miles (173,000 to 168,200 kilometers) from Titan, and from a Sun–Titan–spacecraft, or phase, angle of 56 degrees.

Courtesy NASA/JPL/Space Science Institute.

Image 15 ***Huygens'* separation from *Cassini*:** this artist's conception of the *Cassini* orbiter shows the *Huygens* probe separating to enter Titan's atmosphere.

Courtesy NASA/JPL-Caltech.

Image 16 ***Huygens*' descent:** an artist's rendition of the expedition. *Huygens* is about to reach the surface of Titan, and its parachutes have been deployed to slow its descent.

Courtesy NASA/JPL-Caltech.

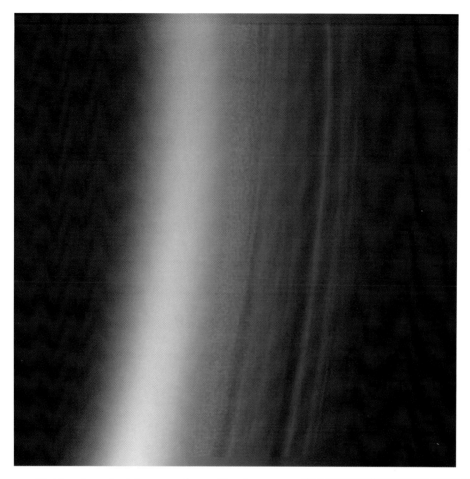

Image 17 **Titan's colorful haze:** above, Titan's upper atmosphere consists of layers of haze extending several hundred kilometers above the surface, as shown in this ultraviolet image of Titan's night side limb, colorized to look like true color. This is a night-side view, with only a thin crescent receiving direct sunlight, but the haze layers are bright from scattered light. The image was taken with *Cassini's* narrow angle camera.

Image 18 **Pebbles on Titan:** right, this image of the surface of Titan was returned on January 14, 2005 by *Huygens*. It has been processed to add reflection spectra data, and gives a better indication of the actual color of the surface. The rocks are pebble-sized, a few inches each across. The surface is darker than originally expected, consisting of a mixture of water and hydrocarbon ice. There is also evidence of erosion at the base of these objects, indicating possible fluvial activity. The image was taken with the descent imager/spectral radiometer.

Both images courtesy NASA/JPL-Caltech.

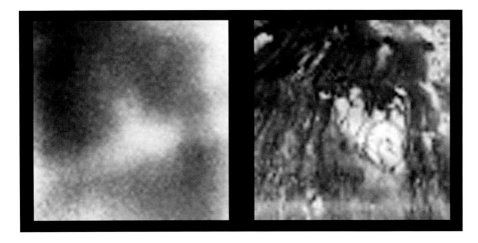

Image 19 **Titan's surface:** These images show the surface of Titan at two different infrared wavelengths. They were captured by the visual and infrared mapping spectrometer on board *Cassini-Huygens* as the spacecraft flew by at an altitude of 1,200 kilometres. The image on the right, taken at a wavelength of 2 microns, reveals complex landforms with sharp boundaries. The image on the left was taken at a wavelength of 1 micron and shows approximately what a digital camera might see.

Both images courtesy ESA.

Image 20 **Titan's Riviera?** This composit was produced from image returned on January 14 2005 by *Huygens* during it successful descent to lan on Titan. It shows th boundary between th lighter-colored uplifte terrain, marked with wha appear to be drainag channels, and darker lowe areas. These images wer taken from an altitude o about 5 miles (8 kilometers The images were taken b the descent imager spectral radiometer.

Courtesy NASA/JPL-Caltech

The composite infrared spectrometer (CIRS) searches for new stratospheric hydrocarbons, measuring oxygen compounds in the stratosphere to determine whether the rings are the source, and performing long observations to analyze the composition of Saturn's atmosphere at different altitudes.

Next, two previously unknown moons are discovered orbiting Saturn between Mimas and Enceladus. All these observations and discoveries are keeping the scientist teams awake at night back on Earth, and yet we are only a few months into the tour. What a wealth of information this exploration is gathering, all the while making pictures and movies that are a delight for the whole world to see.

On August 23, the ground operators successfully complete a "periapsis raise" maneuver. The purpose of this maneuver is not only to raise *Cassini*'s next approach distance to Saturn later in the year by nearly 187,000 miles (300,000 kilometers), but also to keep the spacecraft from passing through the rings and to put *Cassini* on target for its first close encounter with Saturn's moon Titan on October 26, 2004. An exciting encounter it promises to be.

The *Cassini* spacecraft sends back detailed pictures, as well as spectra and radar data, during this Titan fly-by, as it successfully skims the hazy, smoggy atmosphere of Titan, coming within 750 miles (1,200 kilometers) of the surface. The fly-by is the closest that any spacecraft has ever come to Titan – that is of course before *Huygens* lands there early the next year. The pictures, spectra, and radar data reveal a complex, puzzling surface. It is also the first time that the radar instrument is used at Saturn, to make images of Titan's surface and to collect science data. Radar and also VIMS' images of Titan's surfaces appear to be even better than normal camera observations.

Then, on Earth's Christmas Eve 2004, the *Huygens* probe is successfully released from its mother ship, slowly spinning and rapidly accelerating while falling into Titan's gravitational pull. Three weeks later, on January 14, 2005, *Huygens* literally slams into Titan's upper atmosphere at a speed of 14,000 miles (22,400 kilometers) per hour, almost 4 miles (6.4 kilometers) per second. The descent target is over the southern hemisphere, on the day side so that the cameras can get the most of the light. Protected by a thermal shield that has to withstand temperatures of over 21,000 °F (11,600 °C), the probe decelerates by the atmospheric drag within no more than three minutes and – still falling at Mach 1.5, some 100 miles (160 kilometers) per hour – deploys a pilot chute at about 100 miles (160 kilometers) above Titan's surface. At 11.25 Central European time, the Robert C. Byrd Green Bank Telescope (GBT) of the National Radioastronomy Observatory in West Virginia, USA – part of the global network of radio telescopes involved in tracking *Huygens* – detects the probe's "carrier" signal. It is descending!

Only a few seconds later, the pilot chute pulls away the probe's back-cover and the main parachute, 27 feet (8.3 meters) in diameter, deploys to stabilize the probe. The front shield is released and when it is at a safe distance the probe, whose main objective is after all to study Titan's atmosphere, opens its inlet ports and deploys booms to collect the scientific data. The instruments now have direct access to the atmosphere, and conduct initial measurements of its structure, dynamics, and chemistry. On the probe's way down to Titan numerous stunning images of the surface are made by the descent imager and spectral radiometer (DISR), an optical remote sensing instrument with upward and downward looking photometers, visible and infrared spectrometers, and side and down-looking imagers. There is also a sun sensor that measures the spin rate of the probe. The

Doppler wind experiment (DWE) is now busy determining the direction and strength of Titan's atmospheric winds, while the wind-induced motion of the probe is measured from when the main parachute deployed, continuing down to the surface.

After 15 minutes into the descent, at about 75 miles (120 kilometers) above the surface, *Huygens* has slowed down too much so it releases its main parachute while a smaller 9.8 feet (3 meter) drogue chute takes over to allow a faster plunge through the atmosphere. The probe is now coming down at about 165 feet (50 meters) per second, or about at the speed we drive on a highway. In the lower atmosphere, the probe decelerates further to about 16.4 feet (5 meters) per second, while drifting sideways at about 5 feet (1.5 meters) per second, a leisurely walking pace. It is a bumpier ride than expected, the engineers and scientists conclude later, but *Huygens'* descent is going almost exactly according to plan.

The instruments continue their work for over two hours. The gas chromatograph and mass spectrometer (GCMS) is determining the chemical composition of Titan's atmosphere, while the *Huygens* atmospheric structure instrument (HASI) investigates the atmosphere's physical and electrical properties, sensing temperatures and pressures, even listening to electric particles. The aerosol collector and pyrolyser (ACP) has been collecting aerosol samples from the top of the atmosphere down to an altitude of about 40 kilometers and then again in the cloud layer, between altitudes of about 14 miles (23 kilometers) and 10.5 miles (17 kilometers). Aerosol particles are liquid or solid particles uniformly distributed in a finely divided state through a gas, and play an important role in the precipitation process – here a methane rain providing the nuclei upon which condensation and freezing take place. They participate in chemical processes, and influence the electrical properties of the atmosphere.

Huygens begins to emerge from the haze only at 19 miles (30 kilometers) above the surface, like a plane landing through low-hanging clouds.

All this has to work at the same time, and there will not be a second chance. The data are transmitted directly to the *Cassini* orbiter, which is now flying over Titan at an altitude of (at its closest approach) 37,500 miles (60,000 kilometers). Earth-based radio telescopes are also detecting the carrier frequencies directly, which is good because the DWE data, it is discovered later, needs to be collected and analyzed though these observations. It turns out that one of the *Huygens'* two receivers on the *Cassini* is not working, and this experiment was to use the faulty one.

The entire descent lasts about 140 minutes before *Huygens* reaches the surface at a speed of 16 to 20 feet (5 to 6 meters) per second. When the probe lands, it is not a thud or a splash, but a splat; it lands in a Titanian mud of hydrocarbons, a claylike surface saturated with methane. At this time, the surface science package (SSP) and its nine sensor subsystems goes into full operation, with the primary aim of characterizing Titan's surface, after it has already made many useful atmospheric and deceleration measurements during the descent phase. SSP's scientists want to know more about the physical nature and condition of Titan's surface at the landing site, measure the abundances of the surface's major constituents, and determine the thermal, optical, acoustic, and electrical properties of the "mud." Doing all that will also serve up new knowledge for the subsequent interpretation of data collected by the orbiter radar mapper and the other instruments during future Titan fly-by's – interdisciplinary science at work.

A 20 watt landing lamp illuminates the landing site for as long as 15 minutes after touchdown. Spectacular images captured by the DISR instrument from above the surface and by SSP on the ground reveal that Titan has extraordinarily Earth-like meteorology and geology. With some imagination, as many later felt when seeing the first pictures, this could well be called Titan's French Riviera! The pictures show mountains bordering on what seems like a dry sea.

The images show a complex network of narrow drainage channels running from brighter highlands to lower, flatter, dark regions. These channels merge into river systems running into lakebeds featuring offshore "islands" and "shoals" remarkably similar to those on Earth, and also on Mars – a planet that seems to have something in common with Titan's landscape. *Huygens*' data provide strong evidence of liquids flowing on Titan. However, rather than water as on Earth, the fluid involved is methane, a simple organic compound that can exist as a liquid or gas at Titan's sub-340 °F (170 °C) temperatures. Titan's rivers and lakes appear dry, while surface images show small rounded pebbles in dry riverbeds, consistent with a composition of dirty water ice rather than silicate rocks. Yet this ice is rock-like solid at Titan's temperatures.

Titan's soil appears to consist at least in part of precipitated deposits of the organic haze that shrouds the planet. This dark material settles out of the atmosphere. Washed off high elevations by the methane rain, it concentrates at the bottom of drainage channels and riverbeds, contributing to the dark areas seen in images. New, stunning evidence based on finding atmospheric argon 40 indicates that Titan has experienced volcanic activity, generating not lava, as on Earth, but water ice and ammonia. Deceleration and penetration data provided by the SSP instrument indicate that the material beneath the surface's crust has the consistency of loose sand, possibly the result of methane rain falling on the surface over eons, or the wicking of liquids from below towards the surface.

Thus, while many of Earth's familiar geophysical processes occur on Titan, the chemistry involved is quite different. Instead of liquid water, Titan has liquid methane. Instead of silicate rocks, Titan has frozen water ice. Instead of dirt, Titan has hydrocarbon particles settling out of the atmosphere, and instead of lava, Titanian volcanoes spew very cold ice. Titan is an extraordinary world, with Earth-like geophysical processes operating on exotic materials in very alien conditions. The ingredients for an Earth-like evolution are there, the scientists have learned from the recent findings, but it is simply too cold on Titan to create what we have here at Earth. What would happen on Titan if it ever warmed up?

Having survived the descent, the probe continues its characterization of Titan's surface for as long as the batteries can power the instruments and the *Cassini* orbiter is within the beam width of the probe's transmitting antennae. And when *Cassini* slips under the horizon seen from the landing site, 2 hours and 27 minutes after *Huygens* entered the atmosphere, the feast is over. Even before that happens, the *Cassini* orbiter has already reoriented its antenna dish toward Earth to play back the data collected by *Huygens* and to send them to NASA's large radio antenna in Canberra, Australia, where it arrives 67 minutes later. But the Parkes Radio Science station in Australia keeps hearing *Huygens* and the instruments for several hours more, as do sixteen other stations around the world. This is much longer than the engineers had expected *Huygens* to stay alive.

Almost oblivious of all this excitement, the *Cassini* spacecraft continues its mission exploring Saturn and its moons, including over 40 additional Titan fly-bys to explore this Earth-like object. The Saturn exploration has barely begun, and many more revelations about the rings, satellites, and the planet itself are in store.

Cassini's radar spots a giant crater about the size of the Netherlands on Titan as it makes its third close Titan fly-by, exactly one month after the landing. The crater could have been formed when a comet or asteroid tens of miles in size crashed into Titan.

There is the Herschel crater on Saturn's moon Mimas, 81 miles (130 kilometers) wide and with a prominent central peak. Next, the craft passes the moon Enceladus, a good moment for the CDA to take measurements inside the E ring. The CIRS team obtains its first good map of the dark side of Enceladus. Optical remote sensing (ORS) observations and radar scatterometry and radiometry measurements are performed to probe the geologic history and composition of this satellite's surface. And so on ...

The journey continues, with many revelations, and thus press releases, to come. We can only surmise from these accomplishments that this is an integrated science and engineering feat of – why not? – Nobel prize proportions: the precision of getting *Huygens* all the way to that Titan landing site, the equally admirable scientific achievement enabling us Earthlings to take such a solid peek below these mysterious clouds, and the international cohesion to show our sister planet Saturn that we have no borders down here when it matters.

February 2005

6
Crisis and opportunity: how potential failure led to renewal

> *The greater the difficulty the more glory in surmounting it. Skillful pilots gain their reputation from storms and tempests.*
>
> Epictetus (55 AD–135 AD)

A favorite Japanese and Chinese expression much quoted in the Western world these days is that any crisis is also an opportunity. Ancient characters in scripts such as *hanzi* and *kanji* have the purpose of combining two contrasting ideas. There is a Japanese *kanji* for "crisis in which there is an opportunity." Although there is much controversy among linguists about the real meaning of this character (Mair), whole industries of therapists, management gurus, and T-shirt makers have employed it. We however will stay on safer grounds by quoting Greek philosophy.

We asked the *Cassini-Huygens* team about crises in the mission which became opportunities. The Italians had a great struggle with the white paint on their high-gain antenna, and with its multiple functions as a transmitter/receiver, a sunshade, and a shield against cosmic bombardment. Yet they won through on time, to the relief of many on the US side. Those closely involved with the US Air Force and the contractors responsible for the Titan rocket could certainly have considered that building the modified Titan IV version represented a crisis at some times, considering the technical challenges they faced. Dan Goldin even ordered an investigation into whether it would not be better, cheaper, or safer to use the Shuttle for the launch instead, but that idea was later scuttled.

The threats by Goldin and Congress to cancel the whole mission were for many the biggest crisis of all. But here we are less concerned with the politics of funding and financial squeezes, than with direct threats to the mission and its science, and how these were not only overcome, but in fact strengthened the mission's chances of success.

We saw in Chapter 3 that it is never possible to eliminate errors entirely, although it is possible to make extraordinary approximations to perfection. It proved necessary to build self-monitoring and self-repair into the system along with back-up systems. But even these work only in crises the possibility of which has been anticipated, like the spare wheel on a car. There still remain a very small number of mission-threatening occurrences, so improbable that no one thought to guard against them, and these can still ambush any mission.

Often such mistakes are "silly" in their obviousness, like confusing standard metric and English Imperial units (this was the error that caused the Mars climate orbiter, the second probe in NASA's *Mars Surveyor* program, to burn up in Mars' atmosphere), or not polishing the lens of the Hubble telescope in the correct shape. When unforgivable, even ridiculous, errors occur, the system *must improvise an answer*. Such answers often spell renewal. The reason for this is that a mission of this kind is dynamic and extends over long periods of time. This means it is often not possible to undo the error at its source, as it is too late by the time it is discovered: the mission or part embodying the serious error or weakness has moved on. It is necessary to invent and implement a solution on the spot.

What is vital, then, is a power to rally from setback, a mission "immune system" which rapidly mobilizes antibodies to resist the mission being debilitated. In this chapter we examine how three crises that threatened the success of the *Huygens* mission were recognized and dealt with, so that the mission became, if anything, *better than it had been before*.

Although the last crisis was much more threatening that the first two, we shall deal with them in the order in which they were dealt with (which is not the order of their origination). They were:

1   The environmental impact statement, launch nuclear safety approval, and public alarm.
2   The probe's airflow problem.
3   The Doppler shift on the probe relay.

The environmental impact statement, launch nuclear safety approval, and public alarm

The mission team always knew that it was obliged (under US law) to prepare and publicize an environmental impact statement. What it had not counted on was the very high level of public alarm, and the near-impossibility of calming this down.

When your dearest wish is to explore Saturn and Titan, when everything seems to be going for you, and the launch date for this momentous occasion is approaching, who would sit down and imagine the most terrible scenarios of failure, nuclear explosion, the spread of carcinogenic substances across a wide area with hundreds of ensuing fatalities? But if you want a successful mission, or indeed to get approval to launch at all, that is exactly what you have to do. Reed Wilcox, launch approval engineering manager as well as planning, assessment, and integration manager, many others at JPL and the Lewis Research Center (LeRC), as well as technicians at the US Department of Energy (DoE), all had the unenviable job of documenting potentially credible disasters and planning to prevent them.

First, NASA had to publish an environmental impact statement for full disclosure of the potential environmental impact of the *Cassini-Huygens* launch, and this had to withstand public scrutiny.[1] Second, and more importantly, the White House – if not the US President himself – has to approve the launch if nuclear material is being used, and will only do so if all the agencies involved can demonstrate that they have fully assessed the risks involved (which include plutonium release, contamination, and subsequent illness and death).[2] Nor is it only the President who must be informed: the Risk Communication Protocol calls for information to be provided to Congress and the public as well.

Because Saturn is so far from the sun, solar energy cannot be employed to produce the electricity a spacecraft needs. For this reason, *Cassini-Huygens* carried 72 pounds (33 kilograms) of plutonium-238 fuel in its three radioisotope thermoelectric generators (RTGs).[3] This was the largest and potentially the most hazardous cargo of any spacecraft launched to date. It could be claimed that here was an opportunity for NASA and environmentalists to engage in a dialogue about minimizing the risks of exploration, and come up with a solution satisfactory to both sides. After all, we all agree that to endanger Earth's environment would be very foolish, and a nuclear accident would spell ruin for most of those concerned.

However, the DoE and Reed and his team did not have much faith in the environmental groups, and both saw each other as opponents, not partners, in this process. Perhaps in the best of all worlds explorers and environmentalists would find room for one another, but not in this case. Many of the issues were highly technical, and the Coalition for Peace and Justice, a well-organized protest organization that led the Stop *Cassini*! campaign (see the box) fueled much doom-mongering, invective, and conspiracy.

"They've presented numbers that have no basis in fact," theorized the press. If NASA was really a cat's paw of the Pentagon, then polite discussion would not work. The growth of the Internet meant that accusations were fired in the manner of a blunderbuss. Many accusations against NASA and its use of nuclear fuel were based on misconceptions, which there was not the time or occasion to correct. Moreover, those making the accusations were arguably not interested in hearing answers. They were trying to stop

The Stop *Cassini!* campaign

The *Cassini-Huygens* mission stood out as a major target of the anti-nuclear movement. Founded by a Vietnam veteran, Bruce Gagnon, the Florida Coalition for Peace and Justice (FCPJ) became a powerful force in protest at the use of radioisotope thermoelectric generators (RTGs) to power the spacecraft's instruments. The coalition had started much earlier, when Bruce and the physicist Michio Kaku attempted to stop the launch of the first Trident Submarine from Cape Canaveral in 1985. After the *Challenger* accident, and when NASA announced it would use 36 pounds (16.3 kilograms) of plutonium-238 in the *Galileo*'s RTGs, Bruce expanded the organization's presence in Florida by using the press and engaging many local, national, and several global environmental, peace, and anti-nuclear groups and other concerned citizen organizations – ranging from Grandparents for Peace and the Florida Council of Churches to Pax Christi and the Physicians for Social Responsibility – in a protest against the use of the Shuttle to launch *Galileo*.

When NASA announced in 1992 that it would use three RTGs in *Cassini*, Bruce started the Stop *Cassini!* campaign, and the coalition swelled to over 75 different protest groups. Several mishaps in the 1980s – Russia's *Mars 96* and the US *Mars Observer* – fueled the coalition's argument that "NASA does not know what it is doing. Look at *Challenger* and *Mars Observer*, and it is only a matter of time [before a craft with RTGs crash-lands on Earth]."

Through the newly discovered Internet and several rallies in Florida, as well as an active press campaign, the coalition gained strength, not least because of Dr Kaku's reputation and leadership role in publicly rebutting NASA's *Cassini-Huygens* Risk Analysis Report (1997) and Final Environmental Impact Statement. He received national newspaper and television coverage, and even appeared on *60 Minutes*.

The protestors used three acronyms: NIMBY (not in my backyard, which in this case meant, not in Florida); NOPE (not on Planet Earth), and NOSE (not in outer space either). The coalition claimed that NASA's risk assessment regarding the use of plutonium was inadequate and exposed civilians to increased danger, that the technology itself was unethical to use, and that the space program should be stopped if such technologies were the only means available for future space exploration. For the FCPJ the use of plutonium, even the non-weapons 238 type, was unfair, unethical, and immoral. Their worry was not just *Cassini*'s launch, but also the fly-by of the Earth several years later. What if the spacecraft crashed on Earth at that point?

In the months before the launch, even Congressmen joined the fray, and sixteen of them sent letters to the White House asking it to halt the launch. When the launch was near the group held a peaceful protest on October 4, 1997 at the Kennedy Space Center (KSC), in which over 1,500 people participated. But at the actual launch date the press reported that only two individuals had placed themselves close to KSC. Brevard school officials had made plans to delay school at the day of the launch, but *Cassini-Huygens* was off safely before dawn.

The protest briefly revived when the spacecraft made its fly-by within 1,000 miles of Earth. Both the fly-by and the protest were without incident. *Florida Today* reported a protest of 20 people at the Cape Canaveral Air Station, led by Bruce Gagnon, who told the press: "Sooner or later there could be an accident." NASA's risk calculations must have been quite accurate after all, but future missions such as to Jupiter's moon Europa, Pluto, and to Mars are most likely to encounter the coalition again.

the mission, being philosophically opposed to nuclear power and its use in space flight. Nor was NASA entirely effective in its communications with the protesters and press. Even Dan Goldin reportedly was worried that the process could have been handled better, although he did not use it this time to consider cancellation.

Eventually NASA set up its own websites in multiple languages. It gathered all the fears it regarded as legitimate and tried to answer them. Dialogue *was* achieved with some notable critics, namely Steve Aftergood, project director of the Federation of American Scientists in Washington, DC. "He knew enough to ask some very deep, penetrating technical questions," Reed told us. Steve aimed to "bridge the gap" of incomprehension. He was not aiming to stop the mission per se, as most critics were, but trying to find out whether it could be safely conducted. In the end it had his blessing.

In the days before the launch NASA braced itself for litigation aimed at blocking the small launch window. Sure enough it came. The plaintiffs brought a lawsuit in Hawaii. Reed and Beverly Cook regarded this as an attempt to cause major disruption to the planned launch. Key experts had to fly all the way from Florida to Hawaii to assist the Justice Department lawyers representing NASA in responding to the lawsuit. A longer journey within US territory could not have been contrived. Managers, engineers, and lawyers at JPL and NASA HQ worked day and night preparing declarations. The hard work paid off as NASA won its case and the launch got the "go ahead."

Reed returned to Florida, completely exhausted, only to be woken by thunderous knocking on his bedroom door. "Hey, they've filed an appeal!" In a

> Our safety analysis report was two feet thick. It's been thoroughly reviewed. There's absolutely no accident sequence that results in huge amounts of plutonium being released. People have misunderstood the risk. I would not have supported this if my children would not have been safe.
>
> Beverly A. Cook, US Department of Energy

> *60 Minutes* showed me in the program as if NASA and I were indifferent to the issue of nuclear safety. We were not. But I do feel that Kaku and others like him were spreading fears, not facts. Their techniques were very much like walking into a crowded movie theatre and yelling Fire!
>
> Dick Spehalski

last-ditch effort to stop the mission, the anti-nuclear activists filed an appeal before a three-judge Federal panel in San Francisco. It failed, as the judges told the protesters essentially that it was too late, and saw no fault in the process NASA had followed. The mission was on track once more.

Reed had some interesting insights into how judges rule in cases of this kind:

> The nice thing about our legal system is that it operates by rules and the whole Environmental Impact Process has been litigated many times by many different parties and agencies. There have been several cases appealed to the Supreme Court, and the rule of thumb there is pretty straightforward, that if the agency follows its process [that is, it follows its own rules], the courts are not going to substitute their judgment for that of the agency. But if agencies act in an arbitrary or capricious manner, and don't seem to be following their own processes, then the courts will rationalize, "Well if the agency hasn't followed its own process, why should we give any credence to the substance of its argument?"

It would clearly not make sense for judges to rule on questions of scientific nicety, in which they were not qualified. What ultimately matters therefore, are the *dialogues* within NASA (including JPL and participating NASA centers), and within and among all government agencies involved – such as the Department of Defense, the DoE, and the Environmental Protection Agency) – and with interested members of the public who take the opportunity to review and question the substance of NASA's environmental impact assessments. If it can be shown that these occurred and that all substantial issues and potential impacts (such as risks to the public) were carefully considered, then the courts will concur.

In fact, getting the White House to approve a launch is no small feat. There is a lengthy and highly intense process. NASA's scenarios have to be submitted to the DoE for the latter's safety analysis, which then is independently reviewed by an Inter-Agency Nuclear Safety Review Panel, established to coordinate a multi-agency evaluation of the mission's nuclear safety independent from the project. The NASA Administrator eventually sends this evaluation to the White House Office of Science and Technology and its Director. This multi-year process is full of natural tensions, with the panel's experts raising *doubts* to increase their, and ultimately the White House's, *confidence* in the decision to go ahead with the launch.

It is hard to be interested in space and not have an active conscience about the environment. Worlds out in space are born, explode, scatter, and get sucked into black holes. What is dead and empty so far exceeds our own tiny life raft floating in space, that the frailty and wonder of our existence is brought home very clearly to space enthusiasts. So what would anyone involved do if he or she believed there were genuine hazards in the mission, but knew that if he or she was to share these thoughts with the environmentalists, they would use the concerns as ammunition to try to prevent it?

Nevertheless the dialogue worked. With considerable courage considering the general atmosphere in the mid-1990s of "Go, go, go!" Reed and his colleagues sat down and imagined hazards. When you want something that badly, it must be hard to imagine catastrophe, and on a scale that would not only end this mission, but consign NASA and JPL to the history books.

Of course, this is an aspect of scientific discipline. If you wish to be *certain* you need *doubt*. If you want the mission to get to Saturn *safely*, you first need to imagine and so prevent a large number of possible *disasters*. There is no easy way. Reed's team examined over 100 disaster scenarios, many of them worse than their critics could possibly have imagined, and immersed themselves in the horror of such events. Kevin Rudolph and his team at Lockheed Martin Corporation generated simulations of these scenarios. Only then could they render each one extremely unlikely.

With ignorance of the precautions taken by JPL, NASA, and DoE in designing the launch system, spacecraft, and RTGs, one can imagine many doomsday scenarios. The upper stages of the launch vehicle (see the box on page 124) contain liquid hydrogen and oxygen which are notoriously inflammable and explosive. Many thought that if these were to explode, Cape Canaveral and its surroundings might be contaminated by plutonium vapor, which is carcinogenic. However, while any released plutonium could carry hundreds of miles, be inhaled by humans and animals, settle on crops, and enter the food chain, the amounts that could be potentially released in a *Cassini* accident were not sufficient to realize any doomsday scenarios. Here are some of the precautions the team took that precluded these scenarios.

The chances of the spacecraft crashing to Earth, as it orbited, were estimated by JPL at one in a million. And, of course, there is always the chance of a crash on launch. So precautions were taken to limit the amount of fuel that could be released. First, the danger of plutonium depends very much on its form. A capsule of plutonium used in missions like *Cassini* could sit on your bookshelf in your home for 30 years without affecting you. In order for plutonium to be lethal it has to get into the body, either by respiration or by ingestion.

Cassini's plutonium was in an oxide (ceramic) form and encased in iridium. Iridium is used because it is ductile at the high operating temperatures of an RTG, and at the even higher temperatures that can be encountered in re-entry accidents. Were you to put sand inside a balloon and hurl it against a wall, it would deform yet contain the sand within it. This is analogous to iridium covering plutonium.

The capsules of plutonium encased in iridium are placed within several layers of graphite. In fact, the graphite is the same type that is used in the nose cones of missiles that can withstand re-entry into the Earth's atmosphere. RTGs made in this way effectively resist launch accident fires and re-entry heating.

At the very worst an explosion would scatter, not ignite, the plutonium pellets in their double casings. These pellets would be too heavy to be airborne, too large to be inhaled, and too insoluble to be digested. A particle small enough to be vaporized and inhaled is known as a "fine," the size of less than 10 microns. Only a very small portion of an RTG's fuel could ever be reduced to fines in a launch or mission accident.

But the DoE, JPL, and Safety Panel teams were not content to found their confidence upon the known qualities of these materials. They traveled to desert proving grounds, where they burnt, exploded, scattered, and pounded their ingredients in a veritable orgy of destruction. They were prepared to blow up the craft's fuel tanks at the first sign that the launch had aborted, but this did not prove necessary since – as we shall see – larger explosions can actually have an advantage. Tested over 1,000 times in computer simulations, these subsequent "live" tests turned out to reduce the effects of the worst scenarios to a fraction of what had initially been feared. The RTG's containment design was extremely resistant to releasing nuclear fuel in even the most spectacular conflagrations its testers could devise.

All of this constitutes yet another paradox, as Reed seemed to sense as well:

> I don't believe the risks were nuclear, otherwise I would not have had my family down there for the launch. There are more risks from the propellants in the solid rocket boosters, the vapors are toxic in case it blows up. So we only launch if the winds are offshore.
>
> It all went well, and when we went back to the hotel that morning after launch, you finally feel so much less pressure. It helped I had my family there, because I really did go to Disneyworld!
>
> Mark Dahl, NASA HQ Cassini program executive

The basic objective for maximizing nuclear safety is to minimize having a concentration of energy in one place. So ... funny as it sounds, we're using explosions to essentially disperse the propellant and make it safe. Now obviously, when you activate these systems your mission is already over. You are trying to contain the damage.

In fact, the *large* explosions were safer than the *small* ones because they scattered the plutonium pellets further. "That was counter-intuitive," Reed explained. "We gained wisdom with *Cassini*."

The enormous stress and anxiety for the NASA team involved in the launch approval process is hard to put into words here, but this phase of *Cassini-Huygens* was viewed by many on the US side of the team as the most critical moment of the program, as they were not sure approval would actually be obtained until a month before launch.

NASA eventually won the President's approval for launch, on the basis of good preparation in general not just because of this public alarm; the latter forced NASA to go even beyond normal preparation and initial concerns, and outdo those concerns in the horrors it imagined, showing that even the worst contingencies would have their dangers substantially curtailed. The learning loop of the paradoxes inherent in this crisis – *worst scenarios carefully countered*; *elaborate precautions* make for *acceptable risks* – is quite a feat of courage and is shown in Figure 6.1 (overleaf).

In this example the crises were imagined rather than real, and creative responses were called forth, although the stresses of the process and the uncertainty stemming from them constituted a real crisis in their own right. In the two cases following, the crises were very real indeed.

> Earle was the point man with JPL, Lockheed, the White House, the Air Force, all these generals, and the staff of the Energy and Commerce Departments, the Kennedy Space Center guys, and so on. He had a way to bring them all together and play to their strengths. He was a smooth negotiator and somehow got all these instruments in this orchestra to play well together. He had a lot of impact on us finally getting approval.
>
> Wes Huntress on the late Earle Huckins

The probe's airflow problem

A second problem occurred later, six weeks before the October 1997 launch, but was solved very quickly. It affected the launch date by about a week, and launch remained within the "window" in which it had to occur, before crucial parameters would change. Even so, a crisis so close to the launch date with the spacecraft all assembled on the pad was quite a shock to all concerned.

Science missions, as opposed to space spectaculars, like to keep a low profile and get on with their work. But a problem so close to the launch date was bound to cause alarm, and the public relations fall-out was potentially serious, particularly in light of the nuclear protests. Was this not evidence that NASA did not know what it was doing? Was the "dinosaur" about to expire? Was this mission too complicated, too big, and too ambitious for its own good after all?

Once again the issue was seemingly so minor, it appears to have been overlooked. Inside the probe itself, by now mounted on the spacecraft and in the hands of the test team at Kennedy Space Center, were radioisotope heater units, required to keep the instruments warm in the sub-arctic temperatures of Titan. In order to counterbalance the heat generated while

Figure 6.1 Nightmare and awakening

the craft was still in the balmy Florida climate, a specialized air conditioning system was installed to protect the instruments in the probe. Accounts vary, but what appears to have happened was that the air velocity for this system was set too high in the engineering specifications, so when the engineer in charge followed the instructions, the machine fired air too forcefully into the probe. This shredded some internal insulation, raising the concern that insulation debris would contaminate internal components and instruments once the spacecraft was in the vacuum of space. Since the probe was enclosed, none of this was certain, although a test suggested possible problems. *Cassini* was taken off the launch platform, then

Huygens was taken off the orbiter and opened up in the clean room. When the probe was opened up, a degree of damage was evident.

Alas, it was then August 29, and the German team that had designed the thermal system on the probe were all on vacation, their travails supposedly behind them. Ernst Kuliayak, *Huygens* mechanical integration manager and inventor of its unique thermal blankets, was vacationing in Las Vegas, and he was called back to the scene at Cape Canaveral immediately. The rest of his team flew in from Germany, and together with JPL engineers they started on a rescue plan. Ernst remembers not sleeping for the first 48 hours, until the damage was all assessed, then he snatched brief intervals of rest in his trailer. "This was my baby," he told us.

Many were at first doubtful whether the damage could be repaired in time. The initial estimate was three months, which would have delayed the launch indefinitely and embarrassed NASA and its partners. Ernst saw it differently, and worked with his team in three shifts, 24 hours a day, seven days a week. When the insulation had all been cleaned out they found not a single break in the electrical lines, and the entire repair was completed in twelve days. "I felt like a heart surgeon," Ernst confessed. Like a heart, this system had to keep ticking. Everything depended on it, since if it failed, the cold at Titan would crack up the whole spacecraft.

Once again none of our informants blamed anyone for this mishap. "A good engineer always says, 'What are you going to do about that?' never 'Who's at fault?'" Julie Webster explained. None of those we interviewed singled out anyone for blame, although specifying procedures for the test team was JPL's responsibility. The name of the technician who blasted the insulation to pieces was not important to anyone we spoke to.

And yet this incident was crucially different from any other mishaps in that it happened within weeks of the launch, and press and public saw the spacecraft being dismantled again. The issue had gone beyond the influence of scientists and engineers, so someone had to take responsibility. A joint European-American Board of Inquiry was later set up, which focused on the failure to specify what should have been done when the probe grew hot inside. As one informant put it, "Close to a launch everyone is anxious and everyone is looking at you. It was unfortunate timing. No way should the mistake have been made."

> When it happened nobody believed we would reach the launch date. No way, we thought, as postponing the launch for several years was not an option. Hamid was desperate and took responsibility, but he had had no influence on this. Nobody is perfect. It was another victory; a big success. Deadlines have a way.
>
> *Hans Hoffman,* Huygens *integration team leader, Astrium*

JPL's Julie Webster was in charge of closing the ISAs (incident surprise anomalies), a process in which it is necessary to specify everything that

No "Tour de Saturn" without this Titan

October 15, 1997

At Launch Complex 40 of Cape Canaveral's Air Station, the Mobile Service Tower has been retracted from the Titan IV-B/Centaur carrying the *Cassini* spacecraft. The launch vehicle, *Cassini* spacecraft and attached Centaur stage encased in a payload fairing, altogether stand about 183 feet (56 meters) tall; mounted at the base of the launch vehicle are two upgraded solid rocket motors. The Titan IV-B/Centaur launch vehicle consists of a two-stage, liquid-propellant booster rocket with two strap-on solid rocket motors, a Centaur upper stage and 98 feet (30 meter) high payload enclosure or fairing.

The mammoth Titan IV-B is the US's most powerful expandable booster since the final Saturn V placed Skylab in orbit in the early 1970s. This rocket is as tall as a 20-story building and weighs about 2 million pounds (940,000 kilograms) with the solid rocket boosters and fuel. Its 3.4 million pounds (1.54 million kilograms) of thrust is still only half that of its Saturn rocket predecessor, but today it is the strongest rocket Earth has available. By comparison, Europe's Ariane 5 and Russia's Proton rocket have not exceeded about 2.5 million pounds (1.13 million kilograms) and 3.1 million pounds (1.4 million kilograms) of thrust respectively.

On Monday this week the launch was delayed, caused mainly by doubtful weather as well as a minor power problem in the spacecraft. For this launch, weather is unusually critical: unlike the Space Shuttle, which is relatively safe, the Titan IV uses poisonous hydrazine fuel, and safety considerations in the event of a hypothetical launch mishap are very strict – there must be no possibility of debris spreading over populated areas of Florida if the craft has to be destroyed. Opponents considered *Cassini* to be a threat to the world environment if the Titan IV-B booster were to explode at launch (odds of 1 in 450 to 1 in 1,500 according to NASA), or during a planned "fly-by" of Earth in 1999 (a risk of 1 in a million), possibly releasing a toxic and radioactive cloud into the atmosphere. Changing upper altitude winds caused Monday morning's launch to be postponed to minimize any such risks, but now the controllers are satisfied.

The Titan IV-B/Centaur launch vehicle blasts away from Cape Canaveral Air Force base at 04.43 am local time following a flawless countdown. Two minutes and 23 seconds later, the launch sequence continues with the separation of the Titan IV-B booster vehicle, and the Centaur burns for approximately two minutes to place the spacecraft in a "park" orbit. By now, the spacecraft is already at an altitude of over 56 miles (90 kilometers) and travelling at no less than 4,350 miles (7,046 kilometers) per hour.

After 20 minutes in orbit, the Centaur fires for the last time and launches *Cassini* out of Earth's orbit and onto its trajectory toward Venus. The Centaur upper stage separates successfully at 42 minutes and 40 seconds into the flight. Flying now on its own for the first time, the *Cassini-Huygens* spacecraft

> rests for another ten minutes and then successfully opens its communications lines with NASA's Deep Space Network tracking complex near Canberra, Australia.
>
> Two hours later a jubilant Dick Spehalski tells the press that the energy provided to the spacecraft by the launch vehicle was accurate to within one part in 5,000. The angular deviation in the trajectory was described as "insignificant" at better than 0.004 degrees. Mission plans called for an expected adjustment in *Cassini-Huygens'* post-launch trajectory of about 85 feet (26 meters) per second, but flight data shows a mere 3 feet (one meter) per second adjustment will be required in *Cassini-Huygens'* first scheduled trajectory correction maneuver.

happened and had not been anticipated. "It made me feel a bit like a schoolgirl who writes the mistakes her class has made on the blackboard and promises never to do that again. But it does stick in your memory."

This was a relatively minor incident and the remedy was repair, not a creative response per se. Yet, the event stands out because it could well have caused a major delay in the launch the team had worked so hard to achieve on time. The collective response of the JPL and ESTEC's contractor teams to resolve the matter only exemplified how cohesive this international team had become.

Cassini-Huygens launched successfully (see the box above) albeit eight days after the originally planned date because of this mishap, but as it happened, the launch could not have taken place on a better date as the spacecraft could now assume an optimal trajectory and thus saved fuel that now could be allocated to do more science as it got on its way to Saturn.

When we saw Ernst, a man of few words, again in Darmstadt after *Huygens'* successful Titan landing, his eyes were glistering and wet from pride. It was his baby after all.

> In the final minutes of the countdown, when you have reached the point of no return and you have done all you can, you are focused on getting the launch off successfully. But still there are some lonely, quiet moments when you wonder... did we miss anything?
>
> *Dick Spehalski*

The Doppler shift on the probe relay

A very important part of the mission, and from a European standpoint the proudest moment, was the detachment of the *Huygens* probe from the spacecraft's orbiter and its descent by parachute through the heavy orange atmosphere around Titan to land softly on the veiled planet/moon, with its instruments connected by the radio relay link to the antennae of *Cassini*. It led to perhaps the most exciting discoveries of the entire mission.

Yet there can be a problem in communicating between two space vehicle bodies which are both moving relative to each other. In the original plan the relative movement had been calculated at no more than about 3.8 miles per second (6 kilometers per second), leading to what is known as the Doppler shift. Movement interferes with the frequencies used to transmit and receive, as does the changing angle between the two trajectories. A much broader bandwidth than usual is needed as a result. Data will become scrambled and images blurred unless the frequencies of transmission and reception are adjusted.

In early 2000, over two years into the cruise to Saturn, the operations team tested for the first time whether the probe relay would work as expected. As the report (Clausen and Deutsch 2001) said later:

> Unexpected behaviour was observed at sub-carrier [*Huygens*] and data stream level: in particular the receiver [on *Cassini*] performed nominally at zero-Doppler, but showed anomalous behaviour with loss of data when simulated mission Doppler (minus 5.6 km/s) was applied to carrier, sub-carrier and data.

These are technical terms for what in essence spelled total disaster and loss of the *Huygens* mission. Unfortunately the transmitter and receiver had been "hard wired" to save money and to save space, which meant that the system could not be reprogrammed from Earth while in flight – short of sending someone out there, somewhere beyond Venus. It looked very much as if an irrecoverable error had been made, which could well wreck the contribution of ESA to the space mission. Even if the probe was launched and descended on Titan successfully, no decipherable information would be received.

> We have a technical term for what went wrong here; it's called a cock-up.
>
> John Zarnecki, Huygens *scientist*, Open University, England

So what went wrong and how did it happen? There is still no full agreement on this, but what is important is that a solution was eventually found and the mission was saved. "Whose fault?" is not the most urgent question, because in a complex, dynamic system, the problem and the solution are separated by many miles and several years. The priority is to "save the science," *whatever* may have happened to endanger it. Solutions look forward to new discovery, not backward to bygone guilt and oversight.

Nevertheless, now that we have the time and leisure to look back, it is important to consider what went wrong, if only to learn from it. There need not be only one "true" explanation. In any complex system weaknesses may be multiple, and several may contribute to any one breakdown. Several plausible reasons may be better than one, because we learn to take additional precautions. As a result we can consider:

- complex undertakings and elementary errors
- commercial secrecy
- cross-cultural misunderstandings
- mis-specification.

Did all or some of these play a part? Can we prevent their recurrence? It is perhaps best to explain the various sides of the story, and provide some explanations for the parts that may seem less obvious. What follows also includes some conjecture about reasons that things can go wrong in such situations, without any implication that they necessarily applied in this particular instance.

Complex undertakings and elementary errors

The Doppler shift is not exactly rocket science. It is taught in college science courses, and sometimes even in high school. One interviewee likened ignoring it to forgetting to put baking powder in a cake. Someone who did this would fail Cookery 101, in a course in home economics! Nor would this have been the first time that very elementary mistakes have doomed missions. A Mars failure was traceable to confusion between the metric system and the British Imperial measurement system. Foam had been striking the wings of *Columbia* many times before it hit disaster, and the brittleness of O-ring seals on the *Challenger* in low temperatures was well known, and flagged by middle-level engineers, but still went uncorrected.

All these are errors so blatant that they can fly beneath the radar screen of complex reviews and careful testing. Here, then, is another paradox. *Complex work* is vulnerable to *elementary* errors. *Obvious* faults may *escape detection*, being beneath the notice of high-minded experts. You do not lightly ask a master chef whether he has forgotten the baking powder. Peer review may not be at its best when it has to express embarrassing doubts that call competence into question, yet this must happen in order to verify that the right answers are given. And peer reviewers too are not immune from taking elementary aspects of the design for granted.

The initial industrial inquiry board (Huygens Communications Link Enquiry Board 2000) found that ESA's requirements had not included testing the Doppler shift specified at sub-carrier and data stream level.

Commercial secrecy

There is no doubt where the component involved was made, although this does not of itself fix responsibility for the error. Alenia Spazio was the Italian subcontractor to Alcatel, the French prime contractor for the *Huygens* probe. Both were commercial rivals over a number of other space projects. Moreover, Alenia and Italy had spent their own money on this highly sophisticated proprietary communication system. What the prime contractor received was in essence a "black box" with secret electronics inside.

The deliverable was tested at Alenia and again as a receivable at Alcatel, and both times it worked. What appears not to have been tested was whether it would work in the Saturn–Titan scenario, where the differences in relative speeds and trajectories would be considerable. Mark Dahl remembers asking, "Have you tested it end to end?" He was assured it had been tested, but not, it seems, in the right conditions. Simulations usually vary from the real thing. Arguably one reason that the simulation testing was considered to be sufficient was that a similar probe relay system had worked well with *Galileo*.

According to Hans Hoffman, who worked for DASA, a major German contractor, all the other contractors subjected their designs and sub-units to meticulous external review. Possibly the fact that the component was proprietary and contained commercial secrets means that this did not fully occur in this instance, but there is no proof that that was the case. Whatever happened, it is clear that a serious error slipped through a process that should have involved repeated reviews without being detected.

First reactions to this crisis were not encouraging. It took until September of that year, over six months later, for the error to be publicly acknowledged. The ESA team had to admit to this embarrassment in the end. The JPL team's initial reaction was that every effort should be made not to change the mission. Dan Goldin was worried that JPL as a reviewer "might have visibility on this." Alenia admitted rather stiffly, "The circuit operates as designed. The unfortunate part is that it does not work in the mission scenario." It could be said that this was a circumlocution for catastrophe.

Cross-cultural misunderstandings

While it is a truism that errors must be smoked out of the system, and the success of this mission is probably attributable in part to the frequency of reviews, it is extremely undesirable for these reviews to become adversarial. Arguably this was more likely to happen if American experts were consistently searching for European errors. We have seen that Dan Goldin began to demand that reviews be made of *Huygens'* preparations in Europe, in effect claiming the right of Americans to inspect their partners without any reciprocal arrangements. All these reviews found nothing, and had only praise for European engineering, but we now know they failed to detect the Doppler shift problem.

The problem with review processes that could be perceived as adversarial is that the target group has a tendency to perform as if "on stage" before a critical audience. In such cases it views the review panel as a type of court – which of course it is not – and makes no admissions, confesses no doubts, asks for no help, and yields not an inch to its interrogators. In such an atmosphere the target of investigation becomes very reluctant to question itself, and feels nervous about entertaining doubts, with its critics at the door.

As well as the possibility of an adversarial attitude, difficulties can often be caused when communications are between two different cultures, in which different initial assumptions are made, or different attitudes and styles of communication can lead to misinterpretations. Cultural diversity has many advantages, but it does also carry a risk.

Simple translation errors, or even poor choices that are not strictly "wrong," can also cause problems. In several unrelated cases of which we are aware, Italians have interpreted criticism about their work by "foreigners" (or even poorly phrased but not fundamentally critical comments) as personal attacks. We know of some Dutch engineers who tested a bridge, beautifully designed by Italians, for structural faults, and referred to the design as "crazy." This was an idiosyncratic word to use, and probably not a wise one, but it was not intended as a damning insult. However, when one party uses a second language to communicate, the changes of both reading different connotations into a phrase are in fact doubled. The Italians repeated the word among themselves and then walked out of the meeting. We are not, of course, suggesting that Italians cannot distinguish criticism of their work from criticism of them as individuals. Any culture incapable of this would not achieve what Italians have, particularly also in this mission. However, the subtle distinction between criticism of persons and their performances, while quite obvious in Italian, may be lost in translation when speaking a less familiar language. In our experience this particular misunderstanding happens too frequently to be a coincidence, and it may well have played a part in this case.

> Italy's Officine Galileo built another instrument for us, the Star Tracker, the instrument we use for *Cassini*'s precision pointing. It was a real good experience. These Italian engineers are very professional, have incredible machine techniques. And all built to our specs, not like the off-the-shelf stuff we have to buy here. They delivered on time, high quality and trouble free. We had some frustration about the schedule ... as Italians do not have the panic in their eyes we Americans want to see from a contractor.
>
> Chris Jones, JPL

The irony is that Herb Kottler and his team carried out their reviews with a tact and constructiveness that won around many of their subjects. Their review was not only glowing in their tributes, but largely lacking in serious criticism. Similar ESA reviews may have had the same characteristics. But here, too, is a peril. When reviewing someone from a different culture, the reviewer may be so reluctant to cause offence or create misunderstandings that he or she hesitates to make criticisms that could be read as prejudice. This is *critical nonsupport* because the person making the evaluation could be prejudiced against "foreigners." A second form of discrimination, *supportive non-criticism,* is even more common. Here the outsider is not criticized *for fear of this being mistaken for prejudice.*

Notice how these different explanations of what may have gone wrong elucidate and add to one another. In cultures with a direct and specific style of interaction, colleagues find it relatively easy to level with each other. They can be ruthlessly honest and forthright without fear of insult. But in the case of two cultures where values differ, or are simply thought to differ, it is much less easy for a person from one culture to be blunt with one from the other, particularly if that culture tends to be more diffuse. Perhaps a solution here is to find an independent reviewer from the same culture as the principal leader of the review.

Mis-specification

However plausible the reasons suggested above may seem (and however likely they are to occur in different instances), it should be noted that Alenia itself has quite a different explanation for this particular failure. It was difficult to get Roberto Somma, director for studies and technology coordination at Alenia Spazio, off the subject of the antennae of which he was (rightly) very proud. Italian sophistication in communications technology leads the world. When we brought him back to the issue of the Doppler shift, he explained that the relative speeds of the orbiter and the probe were mis-specified in the instructions received from Aerospatiale. Had their numbers been correct, the communication between the probe and the orbiter would have worked.

The real lesson, he argued, was that it was crucial to build operational flexibility into the entire mission so that the spacecraft's instructions could be reprogrammed in flight, as was already possible in a number of other respects. Soft wiring might have cost more and taken up more space on board, but it was still justified. It was such a command data management system that Alenia had originally proposed.

What Roberto told everyone who would listen was that the name-calling must stop and a solution must be found, because everyone would be blamed unless somehow the purpose of the mission could be saved. And he added a footnote, "Galileo had seen the rings of Saturn before Huygens! He had even written it down in an encrypted way to confuse his rival Kepler."

How the mission was saved

One of the more fundamental things about how this problem was solved was that the solution had nothing to do with going back to the roots of the problem. The possible contributory causes were quite irrelevant to the urgent task of saving the science aboard the spacecraft and the probe. Those we interviewed were near-unanimous in pointing out that this was not a "blame culture" but a culture in pursuit of scientific solutions.

Kai Clausen told us:

> In all this we never pointed fingers. The focus was always on the solution. This is a research environment. Mistakes are very common, but it is not our tradition to find anyone guilty. The blame game is a foolish distraction. In any case the person or persons who made the original mistake cannot now do anything about it, so hounding such persons is a waste of precious time.

In a complex, dynamic system, events move on. You cannot roll back the clock even if you want to. What can save the mission is an intervention in quite another place at quite another time. And a solution was found, which for Kai was "the biggest event of the whole project."

In weaker international teams, all this could easily have opened up a rift between Europe and the United States, since the problem originated with European industrial contractors. ESA team member Thierry Blancquaert, while anticipating this, was "shocked and amazed." He was shocked that the error had got through, but amazed by the alacrity with which NASA and JPL in Pasadena set out to remedy the situation. "Their passion was contagious."

Rather than blame the Europeans, JPL shared the responsibility. They had been involved in the reviews of the *Huygens* probe in Europe, and should have detected the error. Several of our European interviewees praised this generosity of spirit and positive attitude. A recovery team was set up and worked around the clock, their morale sustained by the already ailing Earle Huckins, who stayed very close to the group, symbolizing the desire to help of NASA's leadership.

Jean-Pierre Lebreton was impressed by the general reaction to this setback on both sides of the Atlantic. The success story that emerged was a consequence of everyone asking not "Who was to blame?" but "How can we get the science back?" "We never doubted we would find a solution, and when we did the outcome for science was even better as the orbiter now could track the probe longer than designed before," Bob Mitchell told us. "There had to be a way."

Julie Webster, who was in charge of testing all components and systems

---

> The difficulty was not so much the flight geometry, but changing both the flight geometry and keeping the rest of the tour intact. That's where the navigators pulled this rabbit out of the hat. They shortened the first two orbits, added a third, and changed the orbital geometry such that just at the right time, the *Cassini* and *Huygens* would have near-zero Doppler.
>
> The astrodynamics engineer who came up with this was given an Achievement Award, but, you know, we do not really make a big deal out of this around here as it is done all the time. Nevertheless, it was superb work.
>
> *Earl Maize, manager,* Cassini *Spacecraft Operations Office*

during *Cassini*'s integration, pointed out an important lesson of this crisis. A problem originating in Europe was solved in California, not because JPL was "better," but because the solution lay in its area of expertise. There was no way of reprogramming the frequencies used by the transmitter and receiver, but it was still possible to reprogram the trajectories of the orbits around Saturn.

A young mathematics genius, an astrodynamics engineer who had only recently joined the JPL team, redesigned the orbits around Saturn so that the orbiter and probe would fly in parallel at similar speeds, reducing the Doppler shift to near zero, while precious information was transmitted and received. These calculations are formidably complex to most of us, and much of their brilliance lay in altering only the first two orbits around Saturn and then reverting to the original plan, once the probe had completed its journey in January 2005. Had all the orbits been changed, a huge number of pre-planned observations would have had to be redesigned or abandoned, and many sequences would have had to be renegotiated. As it was, the new trajectory actually increased the transmission of science data to a level higher than that envisaged by the original plan. The crisis had indeed spawned an unexpected opportunity to improve upon the initial design. At the cost of little extra fuel, donated by JPL's navigation team, the mission was better than it had been when originally conceived. For example, *Cassini* could now make an early fly-by of Saturn's intriguing moon Iapetus.

Those working on the recovery team, which joined with the navigational team, found it an unforgettable experience, with morale sky-high and the members ecstatic as the solution emerged. Witnesses told us that they had never seen so much concentrated energy in one room at one time. All members recognized the issue. There were a limited number of variables that could still be controlled from Earth. Using these variables only, could a solution be found? It could be, and it was. An engineer told us the moral was that you should always plan to extend any mission, since this gives you additional resources and flexibilities to play with where things go wrong, and allows you to explore new phenomena. The latter may well include phenomena among us here on Earth.

Now that the issue was resolved, the *Huygens* team could go back to worrying about other things, as it did in many ways. Would the parachute hold? Was the heat shield designed with enough margin to tolerate Titan's atmosphere? Were the assumptions about Titan's atmosphere still right? Most of all, would the parachutes be all right after such a long time in storage? Regular things like that were considered, but considered with a much higher degree of intensity and analysis as a result of the crisis.

Qualities emerging from paradox

Crises may still occur even in cases where the majority of potential faults or errors have been eliminated by review, by self-correction, by redundant systems for any fault or error, or by simulations. The source is often some trivial oversight, an honest engineering mistake, or losing track of a scientific law so basic no one dares ask you about it. Crisis brings with it the prospect of *disorder*. What is required is a capacity by the social system for rapid reorganization, to repair the physical system by means of a creative *renewal*. The *Cassini-Huygens* team did just that in this chapter's stories.

It is essential to move on from the place the problem originated, and avoid or at least postpone attributing blame, because the latter may well result after sound analyses and corrective action. Yet, by the time the problem is discovered, the system may have changed, along with ways of healing it.

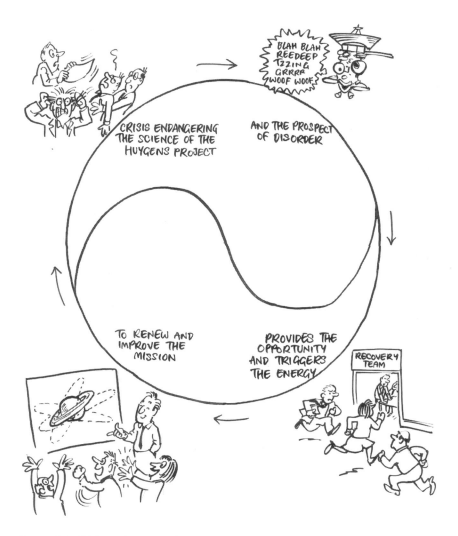

Figure 6.2 Crisis and opportunity

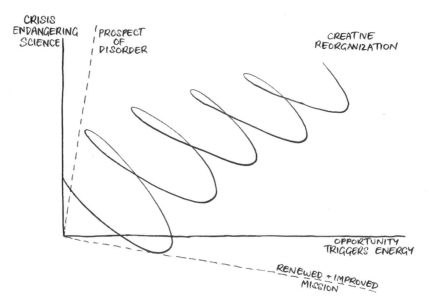

Figure 6.3 Creative reorganization

Crises, real or imagined, spur us to action and provide the *opportunity* for a creative response.

An example of the learning loop associated with these paradoxes (in this case related to the Doppler shift challenge) is shown in Figure 6.2. In general terms, the qualities that have emerged here should be highly relevant for globally networked organizations, with numerous suppliers, outsourced manufacturing, and so on. They can grow in stability and agility by constantly learning through crises and renewal. That this involves a growth dynamic with bigger crises leading to bigger opportunities is made clear in the multi-dimensional illustration, Figure 6.3.

In the case of *Cassini-Huygens*, we doubt that any of the challenges discussed in this chapter would have been met were it not for how this mission revolved around its concept of leadership, one so fitting for the complexity, duration, and cultural environment of this space venture. We take a closer look at this in the next chapter.

7
The equality of elites: leadership and culture

> *Lives of great men still remind us we can make our lives sublime*
> *And in passing leave behind us footprints in the sands of time.*
>   Henry Wadsworth Longfellow (1807–1882)

When it is a question of investigating Saturn and Titan, who knows best? Who should be "in charge" and give the orders? Since the human race has never reached this far before, since this level of complexity is probably unique, since what we *do not know* so far exceeds what we *do know*, who is qualified to issue orders, and why should anyone take these seriously?

The traditional view of authority is that the "author" of the enterprise tells his or her subordinates what he or she wants and intends, and they join in trying to accomplish this. A variation of this is that the person with most experience is assumed to be right most of the time, or at least to have the best chance of being right. You follow her or him because otherwise there would be chaos. Disputes are resolved by invoking her or his judgment, which sides with the most deserving disputant. There are typically timelines for decision making. Without legitimate authority decisions will not be made in time, and the institution must yield to competitive pressures from those who *do* decide. Both hierarchy and elite authority get a bad press, but these tend to be seen as disagreeable necessities. Someone must lead.

This whole question of leadership is made even more difficult when a mission is international. Can Americans give orders to the French, and what if the French decline to heed their orders? Should scientists using a radar instrument tell those using an imaging instrument to subordinate their interests to radar? What is the position of a spacecraft operations engineer relative to a planetary scientist, if one has the means to get to Saturn or Titan, while the other represents the purpose of going there? If either of these is subordinate to the other, the mission could suffer severely.

We cannot give the authority to the "most experienced," to the chief as opposed to the Indians, when virtually every principal investigator is effectively a chief – the most experienced person in his or her own field. What we have is a boatload of chiefs who have given most of their working lives to a particular way of investigating the Saturn system. In exploring such a system it becomes absurd to claim that "surfaces" are more important than "atmospheres" or "magnetic fields." All of these are equally important to an understanding of the whole system, and preferring one to another could lead to conflicts without end and a very one-sided view of a complex, interactive phenomenon. Indeed, if we hope to understand subtle interactions no vital instrument can be neglected, and every viewpoint supplements and complements every other.

With hindsight, we could probably say that one instrument or investigation turned out to be more salient than another. But hindsight we do not have in this case, nor will this chance recur in the working lifetimes of most participants. We do not know which of the multiple viewpoints will provide breakthrough understandings, and because we do not know, we have no reason or legitimacy for the construction of a hierarchy.

The *Cassini-Huygens* team is probably the most dramatic example of multiple elites, with remarkable powers, whose relative value can only be guessed at, and whose claims to command others are completely without justification. To a degree, though, the same situation occurs in thousands of ground-breaking, innovative companies and institutions. To innovate is to lead, and most innovators are not formal leaders. Creative and exploratory persons are made by what they discover. But this happens *after* the discovery, not before. Until we know what they have found, we cannot prefer them. *Cassini-Huygens* is the flagship of highly innovative and exploratory enterprises in a solar system region of mind-boggling complexity.

The theme of this chapter is two more paradoxes. *Elite* persons must be treated *equally*; at least until we have absorbed fully the knowledge they bring to us. This equality does not claim that all approaches are equally valuable, least of all that all approaches are the same. It claims that we do not know which approaches are the most informative until everyone has been heard out. Hence treating them as if they were equal is the best process for discovery.

The second theme is that groups are given direction by both *leaders* and *cultures*: the individual culture of that group, and the wider culture of the disciplines in which its members work. These cultures constitute many centers of excellence, located in various disciplines and informed by different instruments. Effective leaders shape their cultures, and these then shape their membership, so reconciling culture with leadership. Those who led this mission, with a notable exception we consider in the next chapter, did everything in their power to create a culture of *egalitarian elites*. How this was wrought will be discussed under the following heads.

1   Unilateral authority dispersed.
2   The resource provider.
3   Film director and band leader.
4   Defining the culture of the set.
5   Integration and arbitration by team.
6   Selecting for sociability and talent.

Unilateral authority dispersed

At the top of the hierarchy, at least on NASA's side, everything was as usual. As one of the project managers explained to us:

> Goldin said basically, "I want someone's ass on the line." Spe would report directly to NASA HQ and to JPL's director.... Of course, neither NASA nor JPL were responsible for the probe, but NASA had always had a program manager responsible for "everything."

In this case NASA simply pretended that there was direct line authority when there was none. In practice NASA was responsible for the financial, political, and strategic organization on the American side only, while various centers such as JPL and including ESTEC in Europe developed the spacecraft and probe, and numerous institutions and contractors developed the scientific content. All these were connected through contracts and specific interface agreements. There never was one source of authority, Goldin's urge to "kick ass" notwithstanding.

Moreover, as Al Diaz reminded us, there were many issues outside Spe's control:

> The Titan IV launch vehicle was one of them; the RTGs were another.... No one at the top decides what we're going to do on this mission. You cannot define in advance exactly what the mission is going to do and going to be, there are too many uncertainties, too many things that no leader could anticipate.

There were many participants over whom Spe could not have exercised authority at all. Mark Dahl pointed out, "Spe had no authority over the Air Force people, over ESA, over the US Department of Energy. Had he tried to put their asses on the line along with his, it would have been a disaster." In fact, the whole Command and Control ethos only got as far as Dick Spehalski before being rapidly dispersed towards where the knowledge and expertise resided, in every corner of the mission, with every crew member of the *Argo* acting as prince of his own city. As scientist Kevin Baines explained to us:

> Dennis Matson [*Cassini*'s project scientist] does not have the same kind of power you might have imagined 50 years ago with generals and

commanders in Korea. It does not run like that. There is no general issuing orders and no one following those orders. We all manage ourselves by consensus.

It also was that way in Europe. ESTEC's first move had been to send out a "call for mission proposals," which were finally selected by peer review. Ideas for new science missions by their very nature jostle with each other for attention, and join with each other in more elaborate hypotheses. The notion of such ideas being treated unequally is absurd. In Europe the whole mission had been driven by scientific curiosity, which knows no favorites. Romeo de Vidi of Galileo Avionics, Florence, the project manager for the Italian contributions to several science instruments or components of them, reminded us that the Italian unit of discovery was the *officine*, or radically decentralized small workshop of the kind pioneered by Galileo himself. This model still characterizes many of Italy's more creative businesses. In these small units, everyone knows the capacities of colleagues in depth, and creativity is a group phenomenon. And the units appreciate each other.

The resource provider

It might be thought that little remained of traditional leadership under these extraordinary circumstances. If everyone "managed" him or herself and everyone was an authority in her or his field, then leadership in the conventional sense became obsolete. This is not true, however. Leadership was more important than ever, if very much more subtle. In the place of power came influence exercised as an art form, not collected as a tribute. But one aspect of traditional leadership remained almost unaltered, the resource provider.

It was the responsibility of NASA's science leadership, and of John Casani and Dick Spehalski, to obtain, hold, and trim the resources necessary to get a spacecraft to Saturn and Titan, and guide its trajectories once it was there. Every spacecraft development team and every principal investigator or facility instrument team leader was allocated resources, and this time no reserves were held. All science had to make do with what they had, or trade resources with one another. So it was in Europe too, where Hamid Hassan managed the contractors very tightly, with incentives for staying on time and budget.

Al Diaz sees the job of these leaders as choosing the right human resources, giving them the physical resources, and "when they have been given what they need to be successful, get out of their way and let them do their jobs."

However, it is not always that simple. Bob Brown, head of the VIMS team, was given a team of "winners," whose responses to the announcements of opportunity had been the best. Unfortunately, they were nearly all "surface" experts, with almost no "atmosphere" ones. He set about rebalancing his team, using European scientists as a major resource. However, a number of his team members had multiple obligations and failed to show up regularly, and he could do very little about this, save supplement them with the more willing and committed. One member turned up unexpectedly after twelve years of absence!

Although Bob felt his role scarcely amounted to "leadership," he was fiercely protective of his resources. His was a "facility instrument," one wanted by NASA to generate images that would give the public and the science community evidence of the value of the mission. But NASA was loath to elbow aside the ambitions of scientists, so its championship of VIMS was tentative, and contingent on scientists not pre-empting all available resources. We mentioned earlier (page 98) that Bob's instrument was initially qualified so much that he felt a "black hand" hovered over it, execution style. He considered not even accepting the assignment if his chance of being included was not equal to the others, and confronted John Casani insisting the team's resources be guaranteed. John agreed, on condition that he came in at or below the budget of US$30 million. He did, and all was well.

Film director and band leader

We liked the metaphor Bob used to describe the kind of leadership he exercised: that of a movie director or a band leader. A film director cannot tell the actors exactly how to act. Their own interpretations of their parts are crucial. He *can* say "Cut" after the performance meets his standards. He or she is also responsible for the culture of the set, for how well everyone is working with each other; the director sets the scene. A film is shot in fragments, much as the *Cassini-Huygens* mission was an assembly of fragments, and the director's role is to show how each scene fits into the narrative as a whole; analogously, Bob had to show how VIMS was important to the purpose of the mission. He explained, "As a 'director' I have to see that the job is done right. There is no such thing as telling people what to do. I need them to take pride in their work, and swallow their egos."

He had no idea it would be like this when he volunteered to lead:

> I knew nothing of "leadership." I just played it by ear. My own personal style is that of a band leader. I get out there in front and wave my baton in time to the music.... If I had my time again I'd be an Indian, not a chief. I would stick to science, not try to herd cats!

It is clear that an accomplished scientist like Bob feels that what he does is hardly "leadership" at all, and he feels he has sacrificed scientific expertise for something quite nebulous. We would argue, though that he and other leaders are more valuable than they imagine themselves to be.

Defining the culture of the set

We have seen that the film director is the major influence on *the culture of the film set*, reminding everyone of the unfolding narrative and judging how well this is being expressed. He or she defines what this scene or that must accomplish.

The more complex modern industry becomes, the less can any leader possibly know in any detail what should be done. Bob continues:

> The best execution is to make yourself irrelevant.... I don't monkey around with day-to-day operations. I'm a total delegator. My philosophy is to hire really good people and stay out of their way, unless and until they need your help. Unless someone has totally screwed up you do not intervene. This is the only way because, on scientific matters, everybody is Napoleon and they are going to push hard to do the very best they can. It's not true leadership in the strict sense of the word because people only follow when they choose to follow, for whatever length of time they still have confidence in you.

Bob puts himself down, and feels a loss of authority. In our view, he underestimates himself. Leading volunteers is the greatest test of leadership of all, because if you cannot influence them, you fail. Coercing those with their "asses on the line" is no achievement at all, and can doom the entire mission. In fact, as we see from the following excerpt, Dennis Matson also understood this, and he created a culture in which all voices were heard, even those who spoke softly:

> You have to stop some of the more aggressive people maximizing their own results at others' expense. So we set up a sort of "court" where those who might otherwise lose could plead their case. For example, who gets to look at the poles of Saturn, UV or Radio? I know that without my intervention it's going to be Radio because those guys are very outspoken while the UV guys are kind of quiet. So I created this space, where both could present their case in an atmosphere that was friendly and which elicited the entire case, and the UV guys got the nod. I did not favor either of them, but I did ask our panel, "What's the best we can do for *Cassini* as a whole?" and invited them to choose accordingly. How important were these questions to the other questions being posed?

When asked what made project scientists Dennis Matson and Jean-Pierre Lebreton such excellent leaders, Michel Blanc, an interdisciplinary scientist for the mission, answered that:

> they listen to each other and the rest of us. They have no favorites but take evidence. They have strong personalities but do not flaunt these and

always seek agreement. They are accepted as leaders because of the very high level of understanding they show.

Under Matson you did not call for the intervention of the big guns in your discipline, those who might lobby for their own science. Every decision was argued until agreed. Kevin Baines, who spoke liberally to us when we met in Venice and did not mince his words, had memories of the *Galileo* mission, and first-hand experience of science teams who threw their weight around:

> With *Galileo* it used to be, "We're getting this and we don't care what you guys think." We soon knew how this or that team was putting something over on us. Promises would be made and then promptly forgotten. Their instrument was number one and the rest of us could lump it. I remember telling Dennis that he must get the right spirit up front, a spirit in which everyone trusted everyone to negotiate and keep their word, where concessions were repaid. To start the ball rolling I gave up several early observations to get agreement and later observations, and the atmosphere was transformed. Working on *Cassini* after *Galileo* was a form of joy. Of course, there were more trajectories to give many of us what we wanted. Even so, the culture created by leadership was the key.

> Hamid Hassan [*Huygens*' development project manager] was a tough leader towards the ESTEC team, the scientists, and the contractors across Europe. Yet he insisted on frequent communications and good relationships with all, particularly also with the Americans. He always listened extremely well, and connected it to the team's priorities. Wherever and whenever, there was invariably a social event afterwards ... his style made him feared by some, and respected by all of them.
>
> Ben Kroese, Huygens *project control staff member, ESTEC*

This testimony is the key to how leadership of highly complex, highly innovative missions operates. The leader *creates a culture of excellence, and that culture steers the mission*. Further examples of this follow.

Integration and arbitration by team

What constitutes successive levels of leadership is the capacity to *integrate information*. The leader of each unit passes sub-assemblies of information to those above him or her, so that we have a knowledge hierarchy fed by knowledge "heterarchies." The leader does not know more than those who inform him or her, but is responsible for a broader integration of relevant knowledge, and for the fact that it all fits together in a way that advances mission goals.

It follows that the secret of leadership in enterprises of this complexity is furthering expert agreement. This was, and remains, as Mark Dahl put it,

"The largest, most complex interplanetary space mission ever launched." It involved what one of the scientists referred to as multi-dimensional science, "a cat-scan of the planet."

None of this would have been possible without a network of teams, reporting to wider deliberative bodies. Dennis Matson organized the Project Science Group with 28 members, all representative of the investigations involved. There were also overlapping working groups which focused on targets like rings, icy satellites, the magnetosphere, and surfaces, or TOST and SOST (see note 4 page 203). In other words, participants had dual team membership: in their disciplines, and in the targets to which those disciplines aspired. The idea was to discuss how to use different instruments to address common goals and issues. "We were a matrix organization, not hierarchical at all," Dennis Matson explained.

Matson's response to disagreement was to set up teams to arbitrate as scientifically as possible. He created a Special Working Group on Data Volume, and on its recommendation an additional solid-state recorder was installed. Both the *Huygens* Science Working Team and the Orbiter Science Working Team looked at problems arising from interfaces with engineers, and helped to work these through. There were incentives to keep promises. An instrument delivered on time would never be pushed off. It was clear to everyone that the project scientist (Matson) was the equal of the project manager (Spehalski), and that they would agree on all major decisions.

In reaching decisions there had to be the equivalent of a "constitution." Early on in this project, the Joint (American and European) Science Working Group handed down the "objectives for the mission." These became the criteria for judgment when deciding on the merits of two or more conflicting observations.

When recurrent conflicts broke out between the Science Team and the Ground System Team (GST), responsible for operating the spacecraft, Matson created the Ground System Working Group, which arbitrated the dispute. The GST had allegedly been "inflexible" and refused many science requests. It had to explain all grounds for refusal to the working group.

One of the fortunate consequences of these groups and teams was to get science objectives addressed early on, while the scope of the spacecraft was still being defined. If you leave science to a later stage there will not be room for it, and too many decisions will have foreclosed opportunities. Every science team was given three priorities, what it "had to have," "wanted to have," and "would like to have." At least the first two were given high priority.

> We had more than equality between us, more than partnership, we had co-mastery of the design of the whole mission while jointly benefiting from this – scientists and engineers.
>
> *George Scoon, ESTEC* Huygens *design engineer*

Perhaps the most neglected aspect of leadership is the ability to design an entire organization jointly to carry out exciting missions. If you want engineers to respect scientists, if you want disciplines to solve real problems, instruments to complement instruments, then all these levels of expertise need to be designed into the organization from the start. What was achieved was a matrix network of interdependent teams, with mixed membership, so that engineers sat next to scientists from the beginning and never lost sight of their destination.

The pattern is clear: here, leaders systematically favored the integrity involved in voluntary agreements, and the synergies of science, over their own unilateral interventions. Dennis was proud of having to have made only three independent decisions. He wanted science to be the ultimate authority.

Selecting for sociability and talent

Among the chief reasons for the success of this entire mission was the *choice of the right people*. A remarkable number of our interviewees stressed this aspect, on both sides of the ocean. The "right" people were generally the most autonomous, the most talented, and those requiring the least guidance from above. John Casani is especially credited with having chosen very well, especially fellow leaders. A particularly intractable dispute was between the spacecraft operations team and the scientists; Bob Mitchell was appointed and reorganized the interface, and behold, the "intractables" were gone. We would add that choosing the right people at the right time had something to do with mission success as well.

It might be thought that vetoing certain participants was a reversion to old-fashioned, top-down leadership styles, but in fact those eliminated had showed they could not manage their own aspirations, much less reach agreement with others to benefit system science. The persons not included were those who would have necessitated more leadership intervention, not less, if only to settle disputes. John, Spe, Jean-Pierre, and Dennis wanted the disputes settled at the level where they had arisen.

NASA and ESA HQ science leaders such as Fisk, Bonnet, Southwood, and Huntress also played a very subtle, yet very influential role in the selections these leaders made. Key project engineers and scientists needed to be *interpersonally competent*, not just talented. It was not a preference for sociability as a personality trait that was sought, but the hope for a meeting of great minds, although – as logical consequences – these people got along socially very well too (with an occasional exception being pointed out to us quite frequently) and communicated effectively in many ways. You cannot create inter-disciplinary science or science-serving spacecraft if disciplines attack each other, particularly not in a multi-cultural and multi-disciplinary setting.

The people have to be as interdependent as the topics they are studying. "Studying Saturn is like observing a clock," Matson explained. "The whole thing is something beyond the sum of the parts. We must see the essence of the whole system."

In Europe, Jean-Pierre Lebreton made no secret of the fact that he had a "Parliament of Scientists," with principal investigators and interdisciplinary scientists as "senators," and reviews every six months of work just completed. He expected his scientists to forge alliances and affiliations that would illuminate Titan from many angles. Science had a better chance of accomplishing this than politics, because nature was the mediator and final arbiter. At the press conferences after *Huygens'* landing on Titan, these principal investigators complemented each other marvelously, as persons and in their interpretations of what had been observed by their instruments.

In our conversations with other project leaders across the borders, there were many tributes to Earle Huckins, the NASA program director in the crucial development and launch phase, and the man to whom we have dedicated this book. The tributes were given not so much for his technical brilliance, as for how he personified his nickname, "the *Cassini* czar." That may well be a paradox in itself: although also Earle was not directly in charge of any one of *Cassini-Huygens'* constituent parts, he unassumingly wove the relationships with all leaders – from the White House to Air Force generals, from JPL to ASI officials – into mutual understanding and goodwill. He was a "wonderful guy," we were repeatedly told, "committed to good relationships." "He was sick already, but spent much time with us to understand the probe relay problem and help fix it." "He'll be a hero when the list of heroes is published," we were told, confirming what we had suspected all along.

> Earle Huckins helped navigate all the complexities of this project in that period, and everybody knew he was NASA's spiritual leader of the project. He kept Goldin at bay and calmed us all down when the latter had visited us here. Earle was a good buffer between the project and the NASA top. He did a very good job in this period when we needed support the most and felt we did not have it from Dan.
>
> At Earle's premature retirement party we gave him a rock my wife, who liked him dearly too, had found in the backyard and on which she had written "1996 EH1," to symbolize the asteroid we at JPL had decided to name in his honor. We loved Earle and his very soft style.
>
> *Tom Gavin, associate director of flight projects and mission success, JPL*

Qualities emerging from paradox

We have asked, what happens to authority, when investigating novel phenomena by novel means? Should we pick our best people and defer to their greater knowledge? Or should we insist on greater equality among them, so as to ensure that their information rather than their egos prevails?

The answer, as in previous chapters, is that this is a paradox, and that both elitism *and* egalitarianism are required – or more precisely, what is needed are *egalitarian relationships among elite experts*. The answers they seek are not in this discipline or in that one, but somewhere suspended between all of them. All varieties of expertise are not only legitimate but mutually enlightening.

Among history's great explosions of creativity are some very unlikely bedfellows. Who would imagine that a family of bankers would ever befriend artists? It is a very rare connection, even today. Yet, in Florence in the sixteenth century the Medici did precisely that. Michelangelo was an in-house family friend, and so was Botticelli. Even Galileo was for many years protected by the family's influence in Venice, which is where our journey began.

We are not the only authors to note this development. Richard Florida in *The Rise of the Creative Class* (2002) looked at the 15 percent of the US

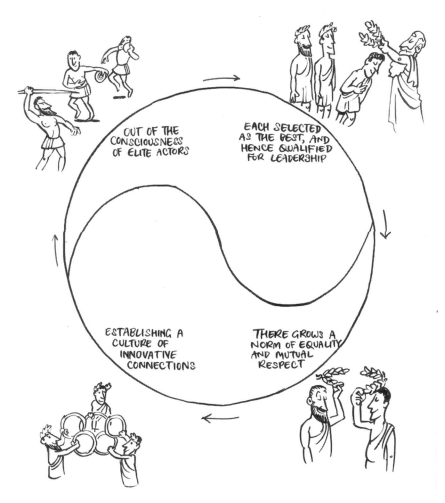

Figure 7.1 Equality of elites

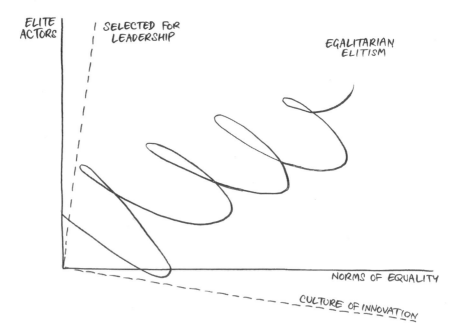

Figure 7.2 Egalitarian elitism

population geographically that produces 85 percent of that nation's innovation. He more recently looked at enclaves of creativity within Europe. What he found in these highly creative sub-cultures was a *plurality of elites*, with no alternative but to negotiate with each other if they wanted mutual understanding. For innovation takes place *between* disciplines, in the new associations that have not been made before because few people have mastered more than one discipline. Out of expertise that achieves mutuality, out of multiple cultures talking to each other, come new discoveries, explorations, innovation, and creations.

The learning loop embodying *elitism–egalitarianism* and *leadership–culture*, in Figure 7.1 portrays the art of leading through the plurality and equality of elites. As before there is no limit to this spiral dynamic. Actors grow more elite *and* more egalitarian, and lead better *through* their cultures getting stronger. See the reconciliation in Figure 7.2.

We continue in the next chapter by showing how not all involved were comfortable with a culture of egalitarian elites, and the complexity implied by this definition of leadership.

8
Simplicity versus complexity: Goldin's cruel dilemma

> *Uneasy lies the head that wears the crown.*
> William Shakespeare (1564–1616)

Up to this point, we have seen paradox after paradox, many of them elegantly reconciled as *Cassini-Huygens* neared its rendezvous. What we have not considered sufficiently is just how painful, difficult, and vexed the fall-out from dilemmas can become. In the previous chapter we saw that only pluralities of elites, each with specialized knowledge, can resolve these knotty issues. None of these brilliant persons was qualified to lead the others or tell them what to do. They had to reach an agreement so that "science could win." Solutions jostled each other until the most effective was adopted.

This process is all very well and good for scientists and engineers immersed in the details of the mission, but what about NASA administrator Daniel S. Goldin, whose term of office extended from 1992 to 2002? Whatever creative styles of leadership were necessary among these pioneers, Goldin was the appointed head of a US government agency, mandated by Congress to carry out its will. The government held him accountable for the command and control of space policy. He could hardly tell Congress and the White House that he had delegated this control to a "Parliament of Scientists and Engineers" from nineteen countries and numerous industries.

If these were conflicts between science and politics, an issue we addressed in Chapter 1, then Dan Goldin was very much the man in the middle, reporting to politicians, yet administering scientists and engineers. He was expected to command what it was in no one person's capacity to achieve unaided. The stress of his job must have been particularly severe because he was placed between two incompatible modes of leadership, one in which "you place your hand on the Bible," as he himself put it to us, and "are directly responsible to three successive Presidents," and one in which science speaks "from a dialogue of diverse experts." The first of these models spells *simplicity*; the second, a daunting *complexity*. The simplicity–complexity dilemma was Dan Goldin's cross, something he could not himself reconcile, but that was reconciled in ways unknown to him by those beneath him.

It is of course necessary to simplify complexity, and *Cassini-Huygens* managed this in its own unique way. Yet Goldin despised and fought this complexity every inch of the way. His dislike of this mission for being too complex was visceral and deep. He told us, "A very wise man told me you have to limit space projects to no more than eight years." Why, we asked him? "Eight years are two Presidential terms."

Well, *Cassini-Huygens* took at least a quarter of a century. There it was on the very verge of triumph, and still Goldin had kept his doubts. We enthused about all the scientists and engineers from so many nations who had cooperated, but Goldin dismissed that:

> When you make things very complex, you drive the costs up, you lengthen the schedule and you reduce the probability of success, and you violate the basic rule of "faster, better, cheaper." The idea is to keep things simple! I have a tendency to resist anything that's too big because the decisions made are on the basis of politics. What you need is a simple relationship because that minimizes the probability of breakdown.

This insistence on keeping everything as simple as possible runs like a continuous thread through Goldin's tenure at NASA. Although we are not uncritical of Goldin – the shortcomings of one man are not the point of this chapter – he struggled, in some respects heroically, with a problem that all leaders of knowledge-intensive enterprises either face already, or will face very soon. What do you do with burgeoning complexity in an American culture that celebrates individualism and pays its business leaders astronomical salaries in the apparent belief that they are "in charge"? No single mind can encompass complexity on this scale, and the attempt to do so places that person under intolerable strain and encourages delusion.

We shall trace Goldin's ordeal through the following topics.

1   An almost impossible job.
2   "Faster, better, cheaper" – a polarized point of view.
3   Command and control.
4   Arch-skeptic and adversarialist.

An almost impossible job

Goldin was appointed by George Bush Sr. in 1992, after the hastened departure of Admiral Truly. The admiral had not "got the message" that space programs were too expensive. He had stubbornly championed bigger appropriations and resisted cuts. Politicians dislike to be "asked" for more money than they are prepared to give. It makes them look mean, and upsets the voters in the districts where NASA contracts have been placed. They expect self-surgery, if not total "hara-kiri," and Len Fisk and JPL wisely started restructuring the *Cassini* project before the administration would demand it. When he joined the fray, Goldin was not going to make the same mistake as his predecessor; during his tenure Congress's appropriations for NASA continued to fall as a percentage of GDP, and in any case he did not approve of NASA's current program when he assumed office.[1]

"Was *Cassini-Huygens* in danger of cancellation?" we asked.

> Absolutely! When I got to NASA, I found a lot of people who thought that just because they were working on something, it was their god-given right to continue. When I got to NASA, the *Galileo* mission was deaf, the Hubble was blind, the weather satellites did not work, and they had spent ten years on the Space Station without any hardware to show for it.... And when I traveled to JPL, the engineers there told me that there was not enough room for innovation.

He had a point there. Nevertheless, JPL's version of this conversation was and is that complex space engineering feats must employ tested techniques, not the most recent innovations. Goldin's words may speak more to his general frustration. He ordered a review of all NASA programs, including *Cassini-Huygens* and the Titan IV/Centaur rocket it would use. So was *Cassini* in danger? We asked again.

> Sure it was! It was on my hit list. Probably the world's finest research organization [NASA] had been reduced to two or three big programs, that everyone fed off. I don't think that they [the *Cassini-Huygens* team] knew how close they came, but I was prepared to recommend to the President that we cancel the project and start another bunch of programs. I wanted them to get the message, but there was a recalcitrance among some of the senior people. They must have thought they knew better than the President. It was their way or the highway.

Meanwhile, in the wider world the original case for large, long-term space programs had begun to look wobbly. In December 1991, the Augustine Report recommended to President Bush a lowering of expectations all around to a level for which the politicians were prepared to pay.[2] "Extravagant" and long-term commitments should be avoided. It was less pay as you go than go as you pay, with scant inspiration to loose purse strings. Then, with the collapse of the Soviet Union happening in the same period – the Evil Empire, as Reagan called it – the rationale for the space program from *Apollo* onward also collapsed, as we have seen. The Bush administration had to find a new justification for the space program, as even Star Wars now looked like overkill.

In June 1991, the community of space scientists took a knock when the House of Representatives – offered a choice between funding space science in general terms and Space Station *Freedom* specifically – chose the latter. It was reported in the media as an anti-NASA vote.[3] Without the umbrella of defense or a visible target, science was once more in trouble. The Augustine Report also stressed the importance of technological and commercial growth, as one might expect in a period of resurgent Republican politics.

That NASA was hostage at the time to a few very large and costly missions was self-evident. The projected development costs of the Space Station had reached over US$17 billion (Wheeler and Siceloff 2001), and Goldin famously refused to look at any more costly redesigns. He enjoyed a triumph of sorts in December 1993 when Shuttle astronauts repaired the Hubble telescope, but that human space flight intervention had cost some US$1.6 billion. Costs were getting out of hand, and Dan was viewed as the "Bold remodeler of a drifting agency" by the *New York Times* (Broad 1993).

Dick Spehalski described to us how Dan Goldin had come into office with a "burr under his butt," and how this made him especially suspicious of *Cassini-Huygens*, against which nothing had so far been proven wrong. An issue that had arisen earlier in Goldin's career at TRW appeared to have some similarity to the rationale behind *Cassini*'s complexity. While at TRW, he had disagreed with Len Fisk, then in charge of Earth Sciences at NASA, about the design of the Earth Observing System – a major science project with a multi-instrument orbiter circling Earth. When the proposed Earth Observing System became too large and expensive, Dan Goldin, on behalf of TRW, reportedly proposed a system of small spacecraft which would put rotating payloads in orbit. While the idea was feasible as an engineering project, it failed, in Len's judgment, as a scientific project. You needed near-simultaneous readings from different instruments to comprehend the Earth's systems. Len threw out the proposal, and Dan never forgave him.

Len Fisk had played an important part in gaining Congressional approval for *Cassini-Huygens*. But now Dan Goldin was in charge, and Len Fisk was an early casualty. NASA was reorganized, and Len was appointed chief scientist, a position without budgetary authority. Len later resigned, and *Cassini-Huygens* had lost one of its ablest protagonists.[4] But the good news was (and this is to Dan Goldin's credit, since he surely must have consented to new appointments) that Len's replacement for space science turned out to be Wes Huntress, who had been involved with *Cassini-Huygens* since the mid-1980s and was known to be a tenacious supporter of the project. He told us:

> The first thing was to save *Cassini*, as Dan wanted to cancel it. You do not say "no" to Dan, so we had to array pro-*Cassini* forces around him. You have to dance with Dan! So we got the White House Office of

Science and Technology Policy [OSTP] involved. "Well, Dan, this is science. Should we be canceling science?" Friends in Congress were approached. "Well, we already approved this mission." We got the Office of Management and the Budget involved. "But we've made a large investment in this mission." OSTP decided to review the project for good measure. Dan got their decision from the White House: the mission was to be reviewed and if it passed, continued. The House Sub-Committee responsible for the space program held a hearing and mention was made about abandoning international partners.

Wes was in no doubt that Dan Goldin would have liked to cancel *Cassini-Huygens*. Others, like Al Diaz (Wes's deputy at NASA) believe that he never really intended to cancel the mission, and that is also Goldin's version of the episode. Asked why in the end he did not cancel a mission he so disliked and publicly called names such as "a dinosaur" and "Battlestar Galactica," he replied: "It would have been unethical. Everyone deserves a chance and that is what I gave them."

But if he was in fact throwing out a challenge, this is not what many others in the program perceived. On Goldin's first visit to Europe, both Roger Bonnet of ESA and Daniel Gautier came away from meetings with him convinced that he was going to cancel the mission. As we have seen in Chapter 1, ESA and the European planetary scientists began to communicate with everyone they knew in the United States, including even a letter from ESA's Director General to Al Gore. Daniel Gautier and others in Europe believed that the European connection saved the mission, and that Goldin did not want the mission because he could not control it unilaterally. He would have to cooperate among equals, and that was not his style.

Goldin's account of his European visit was quite different. He found the Europeans to be particularly "stressed" during his meetings with him. It does not appear to have occurred to him that his attitude towards a project already ten years in planning might have been the source of the stress. They may well have felt that they had a "god-given right" to continue the project.

This European anxiety, we have learned, was exacerbated by the US Congress's annual appropriations process. NASA is "voted" a considerable amount of money for several years for a space program, but will only receive it if this decision is endorsed on an annual basis. As a result of this process any program that first appears to have been generously funded becomes a target for its current and future rivals. It is not unlike affixing a bull's eye to your back so that everybody can take aim. If, in the early 1990s, the Europeans were somewhat alarmed by this annual shooting gallery, they could be forgiven.

Adding to these anxieties and to the difficulties Goldin faced in that early period was that all political priorities had changed. Now it was "The economy, stupid," not the space race. What were programs like *Cassini-Huygens* going to do for the US economy in the remainder of the twentieth century? Not very much, it seemed.

"Faster, better, cheaper" – a polarized point of view

From the beginning of his tenure, Goldin had enunciated publicly his policy of "faster, better, cheaper." What he clearly meant was that smaller and cheaper missions flown more frequently were inherently better. He continued to be proud of this policy when the authors spoke to him in 2003. *Cassini-Huygens* thus became the virtual antithesis of his policy; conceived in 1981, this mission would not pay off for 23 years and more. It would cost US$3.3 billion, and its failure at launch or beyond would have potentially been calamitous for Goldin's NASA, while its gains would be far in the future and would accrue to his successors. In the meantime, a mission that he had inherited, not initiated, consumed a large part of the annual science appropriations. With the overall NASA budget on the decline in the 1990s, he could do a lot more with cheaper and smaller missions, while keeping things simple. He appeared to be a great fan of miniaturization – smaller instrumentation so that cheaper launch vehicles and spacecrafts could carry them.

He also evidently made a mental connection between smallness, cheapness, and innovation. Large size meant caution and obsolete technology. When he got to NASA he found that: "the young people were terrific. They did not know where innovation was going to come from. There was not even a Mars program."

The faster a mission went up, the quicker knowledge would come back, and the more risk could be taken without serious loss. "When I began there were four programs; when I left office there were 25 to 30."

The space science and exploration innovations initiated before and during Goldin's tenure (such as the Discovery program and the New Millennium program) were in several respects admirable. Conceived by Wes Huntress and his boss Len Fisk, the Discovery program was a true innovation in that it promised lower-cost flights, a flight every 18 months, and mission sourcing through open competition, soliciting the best ideas from both inside and outside the agency. These being science missions, the proposer was now to be the scientist who would be held responsible for the science; the mission engineering would be the responsibility of the proposing implementing organization and its project manager. The Discovery program's first two missions were *Mars Pathfinder* and the Near Earth Asteroid Rendezvous (NEAR) mission. Both had been conceived and costed out before Goldin arrived, yet he became a staunch supporter early on, as these innovations played very much into his notion of cheaper, more frequent flights.

The idea of the New Millennium program was to have a technology-driven low-cost flight program to complement the science-driven Discovery program, which would provide flight tests for new technologies.

Mars Pathfinder did a great job for a relatively small investment of some US$300 million – of course aided by the engineering prowess JPL had developed through the *Cassini-Huygens* program. Making a principal investigator directly responsible for the science part of the missions proved popular with some members of the science community, won many over, and widened opportunities. Goldin's championing of Mars as a destination tapped the public mood, combining human and robotic missions, and keeping alive the interest in discovering signs of life.

Goldin's problem was not what he advocated, but what he attacked. *Cassini-Huygens* was perhaps not "faster, better, cheaper," but a single spacecraft with eighteen instruments permits a far better ratio of science to engineering than a single-instrument spaceship with just one or a few science teams. If the purpose is to learn, we have surmised before that *Cassini-Huygens* will be returning more data than any other mission, if not more than many others combined. The risk can indeed be large, and so is the payoff. Particularly for missions to such a distant planet as Saturn, the cost per instrument has to be more favorable in large comprehensive missions than in smaller-size undertakings.

It seems that the issue is in the mental model reflected in Goldin's mantra. That cheap and fast is also *better*, while costly and infrequent is *worse*, is of course a simple generalization, applying to a limited range of targets in our solar system. There was simply no way of getting to Saturn and conducting systems science adequate to the challenge and distance in a cheap or fast way. He disliked this mission because it did not fit his formula.

But the formula itself is anti-paradox and anti-complexity. The idea that any value gets better and better as you polarize it with its opposite is the stuff of Greek tragedy and catastrophe down through the ages. It is, of course, much simpler to pretend that values like cheapness can be extended indefinitely, but it is also absurd. As a cartoon in a recent business journal put it: "John, you have done so well with the limited resources we have given you, that we need you to now do more with none."

Instead of values getting better and better as they near a polarized extreme, what *really* drives superlative performance are clear combinations of opposites. *Cassini-Huygens*' cost per observation is probably the best achieved thus far, and the mission combines cheapness with overall expense. It takes seven years to get there, but once there the data rate is unprecedented, so that *slow* and *fast*, *expensive* and *cheap* are combined. A mission is not worse because it is expensive or risky, but because it ultimately fails.

> I had dinner last night here in Pasadena with the Mars 98 team. I was very candid with them. I told them that in my efforts to empower people, I pushed too hard and in so doing, stretched the system too thin. It wasn't intentional; it wasn't malicious. I believed in the vision ... but it made failure inevitable. They did not fail alone. As the head of NASA, I accept the responsibility. In fact, the system failed them.
>
> Dan Goldin, remarks at JPL, "When the best must do even better"

> For me, "better" meant better for the entire space enterprise as we would have more missions and more science returns, not just better missions as that is a value judgment, and thus a target for detractors. Very few people around me understood that, and the notion was entirely lost with the Mars failures.
>
> Wes Huntress

When the lifetimes of all concerned are on the line, a mission will probably be conducted with extraordinary care.

Goldin approached his assignment with a slogan so simple that it was bound to come back and haunt him. You must, as Scott Fitzgerald warned, be able to hold two opposed ideas in your mind at the same time and still maintain your ability to function. With his formula, Goldin was destined to be only half-right.

Command and control

No issue illustrates Dan Goldin's emphasis on simplicity better than the concept of leadership as command and control. When he first arrived in the early 1990s "every place I looked I saw sloppy work and what I wanted was for people to be sharp." "So you were determined to put the people under fire?" we asked. "Damn right. But I did that with everyone, not just *Cassini*. Come across or I shut you down. I had this reputation of being tough but not irrational." But when we asked what *Cassini-Huygens* had done wrong, he was vague. "Another *Galileo*, another Hubble, another *Mars Observer*, I saw the same trends."

We tried once more to be enthusiastic with him about the international nature of the mission – nineteen nations and many institutions and industries, all working together – but he was having none of it. "No! Too complicated. Three is a crowd and four is a riot. You must keep it simple. There can only be one guy in charge and that was Dick Spehalski."[5] We demurred. How could Spehalski give orders to ESA, ASI, Gautier, Bonnet, the Air Force? There would have been a riot had he tried. Goldin said,

> I held [him] personally responsible. He was accountable to me. When you spread responsibility you bring disaster. On the International Space Station we had three prime contractors; that is an oxymoron. One must be in charge and I held [Casani and] Spehalski responsible. I told them: you go fix it! You cannot spread responsibility.

He showed a lot of admiration for these leaders and their style of management, as well as the wonderful simplicity of the trading system and RecDel, which we described earlier.

What is interesting about this passage is again the mental model. "Spe" (Dick Spehalski) could not tell the international community of engineers and scientists what to do if it was not covered in the very specific agreements on the interface between the various constituencies.[6] He certainly accepted line responsibility from Dan Goldin – even though he worked for JPL, which is merely a contractor with NASA and not a NASA center per se – but Spe was not in a position to demand obedience from the international partners, the US Department of Energy or Department of Defense, or the scientists. Rather, he had to tease the answers from hundreds of minds in a situation where the ultimate authority was not one person, as we have seen in the last chapter, but *in between* a host of leaders. This command and control definition of leadership may have kept Goldin from observing the real paradox-resolving activities of his own subordinates. A project this complex deserved more than a "you are either for us or against us" leadership model.

> There are many lessons to be learned from the Goldin years at NASA. Most are positive, but there are also cautionary lessons, owing in part to Goldin's administrative style. To be the one charged with the radical change of an established agency in complex times is downright daunting.
>
> W. Henry Lambright, professor of political science,
> Syracuse University

Arch-skeptic and adversarialist

We have seen that NASA's administrator at the time sounded as if he was about to scuttle *Cassini-Huygens,* and our conversation with him long after these events seemed to indicate his readiness to do so, but this was not how he explained himself. Rather, he said he had to show he was prepared to cancel the mission in order to give all a wake-up call and to galvanize their efforts towards improvement. He apparently favored an adversarial relationship with his team. People would respond to a challenge if it was tough enough.

Goldin raised the possibility of cancellation widely and repeatedly. Yet he always insisted that this was for everyone's good. He also praised Casani, Spehalski, JPL, the Italians, and the French repeatedly. They were "brilliant," "impressive," "absolutely spectacular," "magnificent," but this was always in response to the challenge he had laid down. The implication is that if they had not been pushed they would not have excelled. We noted that Spe and others were already in control and cutting costs out of the project before Goldin arrived. "I had to get them all to grasp the reality that they were potentially better than the way the project had been performing. These people were outstanding but locked in a time warp. *Cassini* was the only program in development we had." That *Cassini-Huygens* was instead saved by NASA leader Fisk pressing for cost reduction and by the JPL project leaders' skilful de-scoping of the machine has become evident.

"Some of the people who fought me over this had their programs canceled, but John, Spe, et al. knew better and met every milestone, dead on time." The consequences of not understanding Goldin's kind of support were evidently quite severe.

Dan Goldin was nevertheless a convinced internationalist. Telling us of his visit to Italy, he said: "I visited Alenia Spazio and when I looked at their communications system [for the *Cassini* orbiter] it had five different frequencies. Everything they did there was right on the cutting edge. I was overwhelmed with admiration."

Roberto Somma, the director at Alenia responsible for the antennae, remembers the visit a little differently:

> Goldin came here to see the antennae ... it was in the clean room. We had successfully completed the tests a few months before shipment to Pasadena. He was astonished and said to us and later at NASA he could have never imagined finding this level of technology outside the United States. I was not exactly thrilled when Mr Goldin said that.

After 1994 the threats of this mission's cancellation eased,[7] but Goldin continued to demand reviews and ask for new studies, such as seeing whether it would be better to launch *Cassini* from the Space Shuttle. On both sides of the Atlantic, as we have seen, people on the program saw these as attempts to cancel the program by other means. And to be sure, the external review did find concerns around the development of the Titan IV-B rocket to be used, which led to lots of pressure on the contractor and the team to get things back on track. Goldin was right that thorough reviews are for the benefit of all. But if he had not uttered so many doubts about the mission it would have been easier for all involved to believe him.

Why would he not have suggested letting the Europeans and Italians do their own external reviews, inviting Americans to help them as they saw fit? Good reviews require genuine intimacy. "I like *you*, but I will critique your *performance*." It is hard to achieve this fusion of rapport and report across cultures. The problem with the arch-skepticism used by Dan Goldin was that it was so adversarial. As we have seen, you do not get the best results when Italians feel they must put up a good performance before American judges, and the probe's Doppler shift problem may have had its roots in this lack of rapport. Goldin made much of this error – "physics 101" – but the fact remains that the review he had initiated did not catch it.

When qualities do not emerge from paradox

We have seen that Dan Goldin had a job of incredible complexity, not just between politics and space engineering, but between all the dilemmas discussed in this book, which pulled him in opposing directions. He made his position harder by trying to wage war against the complexity of his situation, by what seems to have been chronic over-simplification, and slogans born of a funding crisis in one small interval of time. The irony is that *Cassini-Huygens* will likely be remembered as the crowning achievement of NASA during his term of office. Most of those we spoke to about the role of NASA's leader in this program felt not "tough love" but a breath of ill will. Whatever Dan Goldin intended or really believed, he was never perceived to be fully behind this mission.

When, in 1997, he saw the fully stacked *Cassini* on its work stand in the clean room at JPL for the first time, he reportedly exclaimed, "Holy smoke!" The size and complexity of it all simply amazed him. It was not the reaction of a proud father. A senior JPL person with him responded, "Well, we could not make it bigger, Dan." But he felt like saying: "Damn right, Dan, it is impressive!"

9

Lessons for Planet Earth

> *The spacious firmament on high*
> *With all the blue ethereal sky*
> *And spangled heavens, a shining frame,*
> *Their great Original proclaim ...*
> *Soon as the evening shades prevail*
> *The moon tales up the wondrous tale*
> *And nightly to the listening earth*
> *Repeats the story of her birth ...*
>   Joseph Addison (1672–1719)
>   English essayist, poet, and statesman

So what have we to learn from *Cassini-Huygens*, a mission that seems to have broken all the old rules? In many ways it defies current orthodoxies and points in new directions. So often the search for excellence ends in lists of clichés and stiff rectitude. Yet most of the stereotypes trotted out on these occasions simply will not do.

For example, this was not a unique expression of private enterprise, although the mission could not have been achieved without it. The notion was conceived by academics in national science associations, who then enlisted the support of government agencies. The lead was taken by the non-profit sector, ably supported by many for-profit subcontractors. Any suggestion that excellence of this kind requires persons to seek pure financial profit gets no support at all.

Indeed, even the success of some space agencies seems to have been independent of the performance of their national economies as a whole. Soviet space missions excelled even as the larger economy failed. The joke around Moscow was that Yuri Gagarin was circling the earth but would be back tomorrow, but his wife had gone to buy fish and might not return for days! Many of our interviewees had an abiding respect for Soviet space science. They had accomplished much with far fewer resources.

Nor can it be said that *Cassini-Huygens* was driven by competition. It was no race at all. Europe and the United States were defined as co-equal partners from the beginning and had no rivals. There was no attempt to make one nation more powerful than another, or subjugate and beat another culture. In this it differed markedly from many human and robotic missions, whatever their geographic origin.

Cassini-Huygens was mercifully free of that kind of symbolism. The word derives from the Greek *sym-bol*, "to throw together." *Apollo* and the International Space Station especially "threw together" a number of barely compatible objectives, such as new frontiers, imperial presidencies, world leadership, civilian control of space, astronauts as role models, international cooperation, industrial policy by the back door, science, and engineering prowess. Multiple motives of this kind produce costly mixtures of policy. In contrast, we have seen how *Cassini-Huygens* had a unity that originated in science itself, fusing the engineers and scientists into the achievement of a single, superordinate goal of exploring Saturn's world.

Nor can this mission's success be laid at the door of a single leader or Space Czar stamping his personality on the mission. You can neither command nor control feats of this kind. They rather grow out of wide agreements. Dan Goldin did make attempts at authoritative leadership, but these were deflected and diffused by his direct reports, who referred decisions to where the scientific expertise was located, or where the engineering responsibilities belonged. Very few decisions were forced upon subordinates, and most of these had to do with budgets, schedules, and reviews. Nor were the objections of lower-level engineers and scientists overruled or ignored, as reportedly happened in both the *Challenger* and *Columbia* disasters, with fatal consequences. In engineering and scientific matters any one expert may hold the clues to success or failure. It is neither the rank of such persons, nor their numbers that matter, but the truth of their insights.

In practice, there are multiple rationalities, each bounded and each a partial perspective on a larger whole. All these have different aims yet share the same means of realizing them.

In summary, *external* instructions, incentives, power structures and the like played only minor roles here. The motives to fulfill the spectacular opportunities of this joint mission were *internal* to the mission itself and *intrinsic* to the science and engineering involved. Had this mission failed, the fact that Dan Goldin held Dick Spehalski solely responsible would have been the very least of his and his cohorts' worries! Try wasting 15 to 20 years of your life. Try having the product of your dreams snatched away from you at the last minute by a single, silly mistake like the one of the Doppler shift. Try letting down everyone you have worked with over the years, all the new friends made across so many borders. All of these outcomes were infinitely worse than any sanction anyone could have imposed.

So what lessons can we learn from this? The authors are not space policy professionals, but even if we were, the ramifications of this extraordinary achievement go far beyond this field. We would be remiss if we confined ourselves to recommendations on space policy, particularly international space policy (although we largely agree with the recommendations of the Joint Committee on International Space Programs of the National Research Council and European Science Foundation, made in its 1998 report – see Appendix B).[1] *Cassini-Huygens* could in any case turn out to be a one-time arrangement. Nothing else like this is even in the pipeline. Many national space projects currently planned around the globe are still aimed at "nation boosting," and in some cases have again been whittled down to fit within short-term political objectives. It is once again business as usual, long on rhetoric, short on global science. The danger is that we celebrate the success of *Cassini-Huygens* and forget entirely what made it distinctive, unusual, even a pariah in the higher reaches of the administration. Why did people ever want to prevent this mission?

We suggest that *Cassini-Huygens* has sent Earth five profound lessons that concern us all and transcend space science as such. These are:

1 Mobilizing productive diversity.
2 Leadership shaping culture.
3 Increasing knowledge by integrating values.
4 Meaning as a prime motivator.
5 Learning through paradox.

Let us now look at how *Cassini-Huygens* portrayed these lessons, and what their implications are for the world around us, and more specifically for global business and institutions.

Lesson One: Mobilizing productive diversity

Top performance and novel combinations will almost always be achieved, when bringing highly diverse cultures and disciplines together, by encouraging them to specialize in their favorite activities and strongest suits, while ensuring they share a superordinate goal that commits them to give their best.

It is the combination of these elements that makes diversity so powerful. Simply putting people from different cultures in one room may be "diversity," but it is not necessarily productive diversity, nor will a legal mandate to have diverse people employed lead by itself to that outcome.

Cassini-Huygens is not just looking like being the most fruitful planetary space mission on record; it has also been the most diverse thus far. People and disciplines of nineteen nations and numerous institutions and companies participated. When we started this exploration about the significance of this space mission, we thought that this diverse group of people had shown very effectively how to overcome cross-border or cross-disciplinary cultural differences. Yet what we discovered was much more than that; they had transcended these differences and had made them a source of energy that led to stellar (international) success. And they may have unwittingly provided us all with a framework for anticipating such performance.

We could say that the main diversity here is the one among the mission participants from the United States and Europe. In contrast to the success of transatlantic diversity in *Cassini-Huygens*, however, in general terms the people of the two continents seem to have been drifting apart, such that many on both sides of the ocean seem to feel increasingly that neither side now champions the values that both sides used to share. This is not the time and place to go into all the disagreements that have come between these two blocks over the last ten years or so, but we can try to understand why what they brought to each other in the *Cassini-Huygens* program has – by way of contrast – been such a triumph of cooperation. Can we understand what has happened? Can we somehow make sure it goes on happening?

First let us ask what the United States gave to Europe in this mission. Most of the resources came from the United States. It could have insisted on being the senior party, but on this occasion it did not, thereby bequeathing to Europe what is one of the greatest gifts of the United States to the world – its social egalitarianism, treating everyone as having the same potential, and sharing the gains from mutual knowledge and experience. Again and again we were told that doubts about Americans among the European participants vanished as soon as they were engaged.

Then there are all the clichés about Americans, which are no less true for being clichéd – their optimism, their openness, their stubborn refusal to be discouraged, their faith that seems to be self-fulfilling, and their trust sometimes bordering on the naïve. That all involved are well intentioned was certainly justified in this case. There is the indomitable spirit of adventure, the frontier that never ends, and the sheer stretch of imagination.

Less recognized and perhaps less well known is what Europe brought to the table. Here are people from a continent that, on emerging from the Second World War, had not enjoyed extended peace since Roman times. With quite extraordinary unanimity the leaders of the countries all swore "never again." Since the late 1940s they have built a Community in which war among the original major powers in Europe to achieve hegemony is not just unthreatened but unthinkable, and where a series of mostly "velvet revolutions" have toppled dictatorial regimes to their east, which now seek eagerly to join them.

Europeans were from the beginning determined to make this mission into a further demonstration of European integration and a testimonial to shared knowledge and science, as opposed to a quasi-military posture or spectacular targeting. The high participation of Europeans in this mission, thanks to NASA and higher than their financial contributions perhaps warranted, made certain that this mission was dedicated more to science than to any other end, and even became a cornerstone for Europe-wide educational initiatives.

Europe also brought its own cultural mosaic to this mission. Unlike the American paradigm of a melting pot, Europeans aspire to be French, German, Dutch, Italian, and so on as well as being European. Theirs is a diversity from which unity is forged, not by "melting" but by remaining different. European independence comes by indirection, according to the American writer Jeremy Rifkin (2004), through being embedded in distinct communities.

One reason Europeans and Americans worked so well together in this mission was that both were realizing the dream of their respective continents, be it the American dream of boundless opportunity or the European dream of diverse contributions to a greater whole. People give their best when committed to a superordinate goal which crosses nations, continents, and generations, and which models new and valuable alliances. Both Europe and the United States have legitimate claims to steer the world to a more peaceful future. In other words, if the leaders of our nations across the ocean on both sides started focusing not on the fact that our values differ, but on the fact that there is energy to be created from the differences between those values, maybe we can achieve more together. This energy can be triggered by any looming crisis or dilemma, just as the *Cassini-Huygens* mission's own crises released vast energies. Today's global problems demand a wide diversity of expertise and have as their superordinate goal "survival" itself.

Below the level of political relationships, look at what this means for global companies, for any global organization. For them there are three notable advantages of diversity.

Doing what cultures do best

The first of these is that different cultures tend to excel at the values they hold most dear. Modern corporations increasingly tap into such national strengths. India specializes in services, especially those requiring advanced mathematics, like software design and user services, and processing for Western financial institutions. China specializes in manufacturing, and increasingly in skilled manufacturing. A large percentage of the world's chip-making machinery is located in or around Shanghai. Malaysia is one of the most multilingual nations on Earth, perfect for cross-cultural coordination and customer call centers. Even when cost advantages diminish, this type of global diversity will remain, and may well cease to be called "outsourcing." Global business is moving towards transnationalism, wherein nations combine their special excellences. Hence the automobile of the future may be styled in Italy, its high-performance engine may be German, its safety systems Swedish, its car computer Japanese, its tires French, its guidance and security systems American, its manufacture in China, its PR and advertising British. Complex products increasingly incorporate the finest work of specialized cultures, and try to make rare and innovative combinations among these.

Italian pre-eminence in radio science predates even Marconi, and explains why the entire Italian space agency was a partner in this mission, and why its representation in ESA alone was not enough. The prowess of Italian designers amazed even the skeptical Dan Goldin. The antennae they designed work flawlessly. The data rate of transmission from the probe to the orbiter and from the orbiter to Earth is without precedent for its speed and clarity.

The very name of the mission reminded every one of the French, Italian, and Dutch contributions to the centuries-old study of Saturn. Europe's contribution to planetary science had come of age, and the spread of knowledge has barely begun.

Diversity is not only served by legal compliance, useful though it is to meet equal opportunity goals. Compliance focuses on rules, not on the power of diversity itself. Diversity is achieved with advantages to all, when highly contrasting forms of excellence are located from across the earth and included in one elegant solution. These people are *both* very different *and* very similar. They are different in culture and background, yet similar in their passions for their knowledge and specialty.

Here is a vivid example of this argument. A colleague of ours, Martin Gillo, helped to establish a factory for Advanced Micro Devices in Dresden in East Germany, several years ago. This factory rapidly rose to being the best of the company's facilities in the whole world, including those in the United States, so that work of the highest complexity was sent there. Part of its secret was its diversity. East Germans wanted to show West Germans and Americans what they could do. Making high-grade microchips was an ambitious goal but by no means the only one. This workforce and its managers were determined to restore Saxony to its former glory among German states, determined to form an industrial cluster that would attract other companies to that area – and indeed Siemens and others followed. Their superlative achievement was the fusion of several superordinate goals, including a challenge to the accusation that being under a former communist regime doomed them to inferiority. Gillo, himself German, skillfully used this constellation of goals and stayed on in the region to become a minister in the government of Saxony.

Having fun

A second value of diversity is the fun, the zest, and sheer fascination of understanding someone completely different from you. Interviewee after interviewee told us how enjoyable and stimulating was the mixture of people in their teams; how comparatively boring it would have been to have nothing but Germans, Italians, or French like themselves. These interviewees would have won no prizes for political correctness. A favorite pastime was laughing at national characteristics and teasing people about these. How long and eloquently it took the French to answer the question they had misunderstood. How the British would use subtle hints and circumlocutions instead of blunt criticism, so that no one grasped what they were implying. How you could not confront an Italian without insulting his mother, his family, and his company, in that order. How the Finnish lack of any recognizable facial expression at least qualified them to play poker. How the Portuguese would not drink coffee in the morning, lest they stay awake all day! And how the Americans call everybody by their first name right away as if friends for life.

Of course, no one took such aspersions seriously. No one contended that their international colleagues were anything less than enlightening, no sooner did you get below the surface. Yet cultural differences were seen as a sport, as delightful eccentricities deserving of ready recognition. It is said of really good company, that "hearts are kind and understanding, but tongues are neither." Badinage, or frivolous banter, appears to have been the order of the day, with general merriment when someone reacted as anticipated. This was true for both the Americans and the Europeans. You can tease unmercifully those held in the bonds of one community. Humor, said Arthur Koestler (1964), is when two contrasting logics clash. The second cuts unexpectedly across the progression of the first, and the surprise makes you laugh.

Diversity leads to innovation

The third value of high diversity is that it leads to creativity and innovation. To be creative is to associate ideas or systems previously remote from one another – indeed, so remote that no one else has thought of connecting them, so that creative persons astound the world. It follows that the greater the geographic and cultural diversity, the more likely it becomes that among this vast variety of ideas and techniques will be found a combination that is unique and world-changing. Immigrants have changed our world, partly for this reason. They are strangers in a strange land, with nothing but what they carry between their ears and no social standing to lose. Accordingly, they offer us physical substance, not just the social charm that comes with the fashionably connected.

Innovation through diversity can have some inspiring examples, also in the commercial world, helping bridge the gap between rich and poor. Take the French company Suez Lyonnaise des Eaux.[2] The company was able to put all the utilities of Casablanca into one large pipe, not under roads but beside them, for easy repair and maintenance. Such innovations show there are opportunities across the world. Without ugly poles and lines you get more cell phones, without existing railway lines more bullet trains, without gas stations more use of hydrogen fuel, without power stations more use of sustainable energy sources. Only when these are proven in developing countries will Westerners accept the costs of changing their methods at home. Advancing technology gives all poor countries a potential shortcut to the leading edge. It is very much in all of our own interests to help them – "pure" research with a legacy of compassion.

Cassini-Huygens also innovated its way to success by improvising solutions "on the hoof," inventing new software, new trading systems, new trajectories as the mission unfolded. It corrected its own errors as it flew and calmed the nerves of its own operators. It was a human nervous system hugely amplified and extended into deep space; humanity exploring other planets.

The message *Cassini-Huygens* convincingly sends those who lead global corporations is that to be successful, they must integrate the three advantages of diversity we have mentioned: to let those from different cultures excel at the values and capabilities they hold most dear; to stimulate the fun of being diverse and culturally different as opposed to minimizing it through "policies and non-discrimination rules"; and to connect the seemingly most remote ideas and capabilities to foster creativity. Central headquarters and decentralized units across the globe each play an equally crucial role in making this type of diversity happen, integrating the role and purpose of both. This is genuine transnationalism.

Lesson Two: Leadership shaping culture

Leaders of many disciplines must shape an overall culture of top-level performance that joins together multiple strands of excellence, and that guides the culture's members to the perfecting of the shared enterprise.

There is yet one more quality of inclusion that makes diversity not a danger but a delight. Top performance of the *Cassini-Huygens* caliber is not the work of single leaders, or people en masse, or iron disciplines, or divine discontents, but of *culture*. All experts relevant to the success of the enterprise – however brief their tenure – must leave their marks upon this culture so that their standards are enshrined, and their contributions embedded within a culture of consensus.

What is it that makes upwards of 5,000 highly qualified, highly specialized, very diverse people work together? They do so not because they have been ordered or told to, but because they want to and see the necessity. They are spontaneously self-organized, self-realizing, self-fulfilling, self-ruling (autonomous), and self-critical. Much that is usually outside people has moved within them. They are self-disciplining, and take upon themselves responsibility, not just for their own sub-assemblies, but for how these function in the whole mission.

All this is the work of a corporate or mission culture, which patterns the whole organization and lays down "how we have jointly decided to do things around here." Our view of such cultures is that they fine-tune or reconcile the values of the enterprise.

Probably the finest exposition of the role of culture in society is by Matthew Arnold,[3] Oxford scholar and notable social critic, who wrote in the heyday of imperial Britain. His *Culture and Anarchy*, published as long ago as in 1869, remains a most perceptive insight into what stands between civilization and anarchy. Culture is a "pattern which connects": values with values, people to people, ideas to ideas, and bodies of knowledge to one another. Culture is a major characteristic of teams, shaped by past successes in solving problems. A brilliant mission is thus a fusion of many success stories, carried within team cultures.

In Arnold's vision it resembles a roadmap between places of human habitation. Its members play host to one another, sharing with visitors those environments in which they are most truly at home, and those persons, objects, and pursuits that they hold most dear. We travel mentally or physically from place to place sharing and incorporating experiences. It is culture in Arnold's view that makes the best case for equality between persons, since most of us have had important and incomparable experiences, which need to be weighed equally and appreciated by colleagues. He puts it memorably:

> Culture works differently. It does not try to teach down to the level of inferior classes; it does not try to win them for this or that sect of its own, with ready-made judgments and watchwords. It seeks to do away with classes; to make the best that has been thought and known in the world current everywhere; to make all men live in an atmosphere of sweetness and light, where they may use ideas, as it uses them itself, freely – nourished and not bound by them. This is the social idea and the men of culture are the true apostles of equality.

Arnold was much concerned with reconciling Britain's aristocratic class ("Barbarians") with his own middle class ("Philistines"). He wanted to marry the highest attainments of aristocracy – sensibility, social grace, politeness, and consideration – with the best of middle-class values: fierce individuality, blunt conviction, libertarianism, and action orientation.

We have shown how we believe that this was achieved within *Cassini-Huygens*. Leaders like Spehalski, Hassan, Huttin, Matson, Huckins, Bonnet, Fisk, Huntress, Lebreton, Mitchell, and many others thought continually of how convictions could be orchestrated. Those incapable of harmonizing their own goals with others were pushed back or weeded out in continuous pressure for voluntary agreements. The parliamentary type of "opposition" had to be "loyal" to the mission itself. Europe's "Parliament of Scientists" was constructing a community within a community. Bullying colleagues and assailing them for personal advantage was taboo. "The great men of culture," Arnold insisted,

> are those who had a passion for diffusing, for making prevail, for carrying from one end of society to the other, the best knowledge, the best ideas of their time; who have labored to divest knowledge of all that was harsh, uncouth, difficult, abstract, professional, exclusive; to harmonize it, to make it efficient outside the clique of the cultivated and learned, yet still remaining the best knowledge and thought of the time.

But the characteristic of culture most extolled by Arnold was its search for perfection:

> And because men are all members of one great whole, and the sympathy which is in human nature will not allow one member to be indifferent to the rest or have a perfect welfare independent of the rest, the expansion of our humanity, to suit the idea of perfection which culture forms, must be a general expansion. Perfection, as culture conceives it, is not possible while the individual remains isolated. The individual is required, under pain of being stunted and enfeebled in his own development if he disobeys, to carry others along with him in his march towards perfection, to be continually doing all he can to enlarge and increase the volume of the human stream sweeping thitherward.

As with Matthew Arnold's, the *Cassini-Huygens'* ideal of perfection was made up of many thousands of increments, always improving but never quite arriving, of values that organized each other, with no master abstraction or idealized authority, no altar on which opponents are sacrificed. A spacecraft with many instruments striving for a "perfect" harmony that remains ever elusive, for a complexity never quite complete. This was a feat of successive approximations.

Culture allows a succession of leaders each to leave their mark. We saw earlier that no one is qualified to lead everyone all the time, because there are so many princes of so many domains. What happens instead is that each specialist comes to the fore at that moment when his or her specialty is being considered, and although the leader may lead for but an hour or two, that definition of excellence, those standards of performance become engraved upon the culture. They become the "standards we must attain," "the criteria by which we judge our own attainments." Culture becomes a tablet on which varieties of excellence are writ large, a compendium of expert advice. What registers on the culture and what does not is a question of mutual agreement, so that effective cultures become a record of agreements, minutes of an extended consensus reaching over months and years.

Culture is more than a record of what has been decided. It is a recipe for how we treat each other. We elicit dissenting opinions, listen to each other with patience and respect. We inquire as to what additional information is required to settle disputes. We put a sanction on rudeness, interruption, bullying, hectoring, and pressure tactics. We appoint teams to address simmering disputes and take their recommendations seriously. We invite those who dissent to tell us what precautions are necessary, what alternatives they propose. We reciprocate for kindness and concessions. We ask, "What is best for the enterprise as a whole?" not "What is best for me?" We admit we do not understand a question. We follow through on all requests for action, and so on. In other words, culture creates and maintains social norms which guide the social process – critical supportiveness, reviewed designs, friendly competition, loyal dissent, doubted convictions, and so on, all in search of the optimal synthesis searching for perfection.

A company that makes excellent use of culture is the British United Provident Association (BUPA), a private health insurer and provider. Its notion of "training" its call-centre staff is to mine the experience of its own people in helping their own family members confront illness. When customers call in, having paid their premiums for many years, they rightly expect very prompt attention to any medical emergency. What they usually get at the other end of the telephone line is someone whose father also has Alzheimer's disease, or whose mother had a mastectomy or heart bypass operation, or who has nursed a child through leukemia. The understanding will be deep and poignant, the advice very personal, the attachment real. Moreover the call-taker is highly motivated to speed medical assistance on its way and take personal charge of the case. In this instance taking an inventory of cultural resources and matching these to customers' needs is the best strategy of all. Development consists of staff telling fellow workers of their ordeals as carers, so the team can discover who has survived what, and thus nurture the one life we all have (Trompenaars and Hampden-Turner 2001).

So culture now *mediates between* leaders and subordinates. It is an agreed and evolving set of norms, values, and procedures, bearing the marks of many experts, to which members have given voluntary assent. It serves as a continuing consensus governing discourse and decision making. Organizational complexity of the likes of *Cassini-Huygens* is increasingly the rule not the exception, particularly in this rapidly globalizing and connected world. Again, it also shapes the role of our political leaders differently – it is not "my way or the highway," but rather that all political entities, regardless of their relative economic or military strength, share the responsibility of creating a culture within which we can safely and harmoniously explore solutions.

Finding a solution for our global environmental threat could, for example, be the shared standard of excellence and perfection, while it might be adopted as a global social standard that protecting polluting industries or products is no longer acceptable behavior. As it concerns the global enterprise, the essential role of leadership could in effect become one of establishing a corporate culture that transcends individual cultures and binds highly diverse business components and individuals. That can hardly be the same national or corporate culture as that of the global headquarters. The *Cassini-Huygens* mission has clearly shown that Jim Collins (2001) was quite right that great leaders combine a personal humility with professional will. Great leaders demonstrate a compelling modesty, and are never boastful, since they internalize standards into the culture of the organization, not in the self. They look in the mirror for responsibility and out of the window to apportion credit for the success of the organization.

Lesson Three: Increasing knowledge by integrating values

When highly contrasting bodies of knowledge, each seemingly "true" yet contradictory, are both respected, both considered on their merits and accorded justice, there is a good chance that they will spontaneously combine into an ever-strengthening, ever-flowing combination of vitality.

One of the oldest ideas in science is that Creation is unified, that reality is somehow integral, elegant, aesthetic, harmonious, whole, beautiful, and true. To many this is also religious doctrine. God had made the universe and His works are the perfect rendition of a divine being. People show their piety by setting out to explore this unity. While divine creation finds little support among secular scientists, the notion of a unified natural order has survived among them.

With or without the belief in a universal God, do we live in a wondrously ordered universe in which all elements will be found to combine into a lawful system, provided we search hard and long enough and keep faith? We use the word "faith" advisedly, because the fusion of truths into a pattern that is beautiful has never been verified and never can be. It is a paradigm (a search pattern) and an epistemology (a path to knowing). Were we to look simply at the millions of scientific observations that will be made in this mission, for example, we could conclude that the data are arbitrary, capricious, random, void, and meaningless. What Einstein meant when he said, "God does not play dice with the universe," is that the order is there and we must look for it.

The belief that we are in search of new unities, still hidden from us, is vital to understanding the motives of the *Cassini-Huygens* mission's members. From the beginning they conceived of this mission as *systems science*, and the Saturnine system as a meaningful whole in dynamic equilibrium. This is less discovered than assumed, and this is currently being verified up there, almost a billion miles (1.5 billion kilometers) away. Are the members of this mission creative in the sense of generating some new phenomena, not there before, or are they simply discoverers of something that has already been there for millions of years?

As so often before in this book, the answer has to be both. People can discover oil beneath their garden and grow rich from what they have discovered, but more complex discoveries which use instruments to gauge near-invisible constructs require the investigator or engineer to have in his/her mind a model of the reality "out there," first created and then found to fit all known observations.

The physicist David Bohm (2002) has argued for an "implicate order" lying behind all science, an assumption of unity, not imposed, but endlessly rediscovered in the work of scientists. This wholeness is not just present at Saturn as a system, it is paralleled in the minds of creative scientists and engineers. It is present in relationships between persons and cultures, within teams and within the mission as a whole. Its product is a unity of knowledge.

It is also present in the contrasting values like *doubt* and *certainty, loyalty* and *dissent, error* and *correction, play* and *seriousness, competing* and *cooperating,* which in chapter after chapter of this book have been reconciled into virtuous circles and connected spirals, suspiciously similar to the DNA helix, the template of life itself.[4]

One of the qualities of the unity for which we are searching is life itself. There is unity without life, but some forms of unity generate life. In fact, life is not guaranteed by any list of chemicals, just as we know there is no life in the primal soup of Titan. A human body deceased a few seconds ago may not yet have lost its constituents, but it does at once lose its organization. Human vitality is a form of dynamic interaction between constituents, including values. To clarify by way of example, the difference between the *Cassini-Huygens* space mission and – and what may seem a surprising contrast – the desire for some countries to resist immigration lies not in values as such (here, *diversity* in tension with *inclusion*) but in the patterns of excitation *between* these values. In this mission, there was the widest of the diversities from which the best contributions from the best minds were selected and included, no matter how strange and unfamiliar. In a country resisting "foreigners," the degree of diversity creates such anxiety and panic that those seen as most diverse are systematically isolated or kept out. This results in an ever-narrowing inclusion, as the two values fight each other in frenzied antagonism.

In each case both values are present, both are strong, both arouse passion, both generate fierce dedication and energy. Yet the outcome of one is a successful exploration of the heavens and the outcome of the other is a society in frenzy. This difference, which makes all the difference, remains curiously elusive and difficult to pin down.

Stellar accomplishments can only come about by intense acts of faith and confidence in your mission. Part of that faith is that – no matter how fragmentary, contradictory, chaotic, disorderly, and diverse the data look – values such as unity, order, harmony, balance, integrity, and order *can* be found. There are total solutions, whether for a company's customer, or for humanity, or for intervening entities in general. We may feel isolated, lonely, and confused at times, yet there is a place where paths converge, where individual perspectives prove complementary, where the model created in our mind's eye fits the external reality, so there is both creation and discovery.

A well-known company that has always sensed the capacity of its own culture to achieve unity is Hewlett-Packard (HP). Dave Packard and Bill Hewlett originated the practice of "managing by wandering around," a process made famous in Peters and Waterman's best-seller *In Search of Excellence* (1982). Managers find out where their company should be heading and what it should do by "wandering around" among their best people and keenest minds, and asking for their dreams, aspirations, and thoughts. Such a procedure would lead to chaos and unstoppable divergence but for the hidden integrity that lies behind the quests of science. For many years in the last century, HP innovated brilliantly by moving in a direction that its best minds had indicated. This was less a strategy than a strategic direction, which sensed where lay the secret gardens of low-hanging fruit. The trick is to work with the unifying tendencies of the human mind.

Lesson Four: Meaning as a prime motivator

Perhaps the most powerful motivation and impulse in humankind is the search for wider meanings. New meanings emerge when tasks are performed for a purpose beyond themselves, to the extent that these help make sense of other phenomena, and bridge the gaps between people, disciplines, cultures, and generations.

We have seen that the unity of science, the fusion of America and European dreams, the connection between diverse ideas and the capacity of cultures to bring values into optimal relations, all contribute to the process of perfecting shared tasks. What all these have in common is the search for meaning, which is especially strong when people face a great void of ignorance. It was Viktor Frankl (1959) who, when sent to a concentration camp, confronted a scene that was meaningless, horrific, absurd, and mindless. Paradoxically, he discovered within himself and a small circle of companions the capacity to create those meanings so absent from his environment, and this became the source of their survival, while all around them perished.

Meaning comes about through *connection* – two, three, or more pieces fitting together and suddenly becoming a new coherent reality. With *Cassini-Huygens*, literally thousands of observations are parts of a greater meaning inherent in the whole. Looking for whole meanings and connecting these to expand consciousness is how the human personality develops. It was William James (1949) who described the moral person in the moral culture:

> He knows he must vote always for the richer universe, for the good which is most organizable, most fit to enter into complex combinations, more apt to be a member of a more inclusive whole.

Sometimes the capacity to relate and organize brings ecstasy. One of the greatest moments in theatre and cinema occurred in *The Miracle Worker*, starring Ann Bancroft and Patty Duke. A teacher (Anne Sullivan) is attempting to educate a deaf and blind girl (Helen Keller). All she has to communicate with are the child's hands and fingers, upon which she makes impressions corresponding to the people and objects in that child's small tangible world. The child regards the finger playing as a game, much preferable to sitting alone in the dark and the silence, but a game nonetheless. She has fits of temper when the "game" goes on longer than she likes. But then quite suddenly she discovers that the impressions on her fingers correspond with features of her environment. The water she felt flowing over her open hands comes from the pump. She has been taught a language. Not only does the sign for water and pump correspond with a reality she can feel, but one phenomenon derives from another. Her world is making sense! She runs around the room in a fever of excitement "signing" the objects she touches and finally signing the face of her teacher and embracing her. Rarely has our inherent thirst for meaning been given more vivid expression.

We suspect that there was a very similar form of excitement when Saturn loomed out of the dark of two millennia and Titan's veil was at last penetrated. Then thousands of isolated observations converged upon one meaningful whole, as disciplines talked to disciplines and patterns began to connect.

Can we not employ similar fusions of human passions to explore the knots and puzzles of our world and businesses? In chapter after chapter, we used learning loops to connect what many people see as different or as even "opposed" viewpoints. It is the creative connection between these seeming oppositions, recognition of, respect for, reconciliation of, and finally realization of such differences, that provides new, exciting meanings. If a new form of safety harness for young children in automobiles means that annually roughly 2,000 will not be flung through windshields in a collision, then technical prowess plus social value have created new meanings, with the capacity to give their precious lives a new direction and purpose.

Many activities that look at first glance humble and prosaic may be neither when we trace their full ramifications. We have already mentioned Suez Lyonnaise des Eaux. It happens to enjoy a more than 50 percent share in the world market of foreign-operated water utilities, not bad for one company in one nation. How did it achieve this? First it gave the seemingly utilitarian task of building sewage systems the importance this deserves. Dramatic rises in life expectancy owe relatively little to wonder drugs or super surgery. What raises life expectancy from around 40 years to well over 60 is drinking clean water and not living amidst your own wastes.

The company is thus not only transforming public health in poor communities worldwide, it operates a "post-colonial" strategy, using French embassies abroad to sell the vision. It believes that every country should own its own infrastructure. It therefore secures 20-year concessions to rebuild and render state of the art a city or country's water and sanitation system, even training an indigenous workforce to operate and maintain it. The system is then handed back to national or municipal ownership. So even relatively "low-tech" solutions can have wide social ramifications and broad meanings for the challenge of human development and public health.

What *Cassini-Huygens* tells us implicitly is that great leaders and great companies are Searchers in Chief who lead this quest for meaning. They help us to transcend our own mortality and leave a legacy that can last as long as the human mind.

Lesson Five: Learning through paradox

Learning that expands awareness and generates human knowledge is at its best when it is the consequence of a true transformation, when values suddenly cease to conflict with one another and become synergistic and reconciled.

We began this book by asking how leaders "managed values" in such a way as to achieve excellence. We argued that values are differences, and hence they assume the form of paradoxes or dilemmas. These dilemmas will eventually dismember us mentally and socially unless we learn how to reconcile them. Reconciliation is the work of culture. The time has come to take stock of this argument. How did it fare? Throughout this book we followed the thread of this argument.

Much has already been said about the nature of these bifurcations and paradoxes. Here we will concentrate on how they connect and help us learn from one another. On the face of it, dilemmas and paradoxes seem to present very unfertile ground. How many times have we been plunged into boredom or anxiety because the challenges we faced were either too much or too little? How many children play truant from school, playing when they should be serious, erring repeatedly and not correcting their conduct until it is essentially too late for them? Or, at a higher level of abstraction, why did we fight a cold war for nearly half a century, if not to prove that the ideal of cooperation was a false prophet and the ideal of competition a true God?

Yet a major message of this book – no, of the very *Cassini-Huygens* mission – is that such rival belief systems *share the truth between them*. New syntheses are forged out of a seeming clash of values. For example, an immigrant comes to town. He is strange to us, a bit frightening, foreign, "a curious person" as was said of Socrates by his accusers. Surely no one so far from our settled opinions will readily succeed among us! In that case, why was one-third of Silicon Valley's new wealth, as of 1999, generated by Indian and Chinese immigrants who arrived in the United States less than 30 years ago (Saxenian 1999)?

There has to be some mysterious middle ground between distance and relationship, so that while we usually find it difficult to relate to persons distant from us, there arise a sufficient number of "ecstatic moments" to generate billions of dollars in new value; moments when this prior distance makes relationships more rewarding, enlightening, and valuable.

Our position is that these sudden syntheses apply to contrasting values in general. Take Johnson and Johnson, a global supplier to hospitals and consumers (Collins and Porras 1994). Arguably no company has served its shareholders so well for so long. Yet in its credo, as we have seen in an earlier chapter, it announces, "Shareholders come last." How come that persons put "last" are so well treated? Of course, "last" means last in time. First managers must enthuse employees, then employees must delight customers, then customers buy more products of greater value, and only then can shareholders benefit. The dynamic is circular, like the diagrams at the ends of the chapters in this book. The credo of the company was tested when a rogue shopper inserted poison in bottles of Tylenol. The company knew what it had to do – serve customers first and shareholders through the gratitude of customers. Every bottle was withdrawn, doubtless at ruinous expense. Within two years customers had increased Tylenol's market share.

An inspiration came suddenly to the American anthropologist Ruth Benedict, who in *Patterns of Culture* (1958) tried to explain why two native American tribes were relatively prosperous and happy, while three were utterly miserable, addicted, intoxicated, and despairing. For many months she failed to find a single variable that separated happiness from misery. Then she decided to look not *at* these values but *between* them. All five tribes were very eloquent on the subject of *self-interest* versus *altruism*, moralizing a lot whether they were joyful or woeful. What characterized the prospering and happy tribes was that the difference between egoism and altruism had been transcended. When someone clearly acted in an altruistic way his or her neighbors promptly reciprocated, so that self-sacrifice was avoided and favors were exchanged. In contrast, anyone trying to help another in the miserable tribes was ferociously exploited.

Benedict coined the word "synergy" (as noted earlier, Greek for "work with") to show that when contrasting values work with each other, social benefits result. Following her death, Abraham Maslow championed this insight. He studied a number of outstanding Americans, who had given much to their society. What had they in common? He summarized his findings unforgettably in *Motivation and Personality* (1954):

> The age-old opposition between heart and head, reason and instinct, or cognition and conation[5] were seen to disappear in healthy people where they became synergic rather than antagonists, where the conflict between them disappears because they ... point to the same conclusion.
>
> The dichotomy between selfishness and unselfishness disappears ... because in principle every act is both selfish and unselfish. Our subjects are simultaneously very spiritual and very pagan and sensual. Duty cannot be contrasted with pleasure or work with play where duty is pleasure, when work is play.... A thousand philosophical dilemmas are discovered to have more than two horns, or paradoxically, no horns at all.

The quite extraordinary success of *Cassini-Huygens* has one final lesson for us all. The stellar performance of this mission required a quality increasingly absent from modern life, the satisfaction of the human spirit. This is not necessarily religious, although several religions nurture the spirit. By spirit and spiritual we mean literally *spiritus* (the word is from Latin "to breathe"). Values, especially the softer values, have the capacity to breathe into other values, even the hardest. It is similar to the "kiss of life" that can put breath into an almost drowned person to get his or her lungs moving once again. Ideals can breathe into reality to change it. The self to which we aspire can become actual. The human spirit can animate the dead material universe. Loving support between people can "breathe into" criticism to render this constructive and mission-sustaining. Inclusion can breathe into the potential Babel of diversity to create a colorful, dynamic culture, capable of great feats. Values which qualify values to create new forms are the very breath of life.

We can think of no better example of this human spirit than the one we experienced with the *Cassini-Huygens* team at the climax of this mission: the subject of our Epilogue, next and last.

Epilogue

Tensions, tears, and flow

January 14, 2005
Darmstadt, Germany
Seven years and three months after launch at Cape Canaveral ... and six months and two weeks after the spacecraft's impeccable insertion into Saturn's orbit, entering right through its rings and skillfully avoiding potentially fatal debris. All systems and instruments continue to operate "nominally," as the weekly mission reports keep telling us.

We are in the European Space Operations Centre in Darmstadt, in a big hall filled with people from many nations, including those who were at the launch, come together again. I find myself saying "us" as I truly feel that I have become a crew member on this discovery ship to Saturn. I recognize many of the faces and names, the ESA, NASA, ASI, and contractors' employees we have met along our journey; the scientists, the Huygens engineers, their family. I see Hamid's daughter, Jean-Pierre's wife, Don Kindt who has come on his own expense to see his old friends from Europe. There is Spe who has come from Florida where he now plays golf almost daily, and of course there are Daniel and Toby, and so on. And also Ron Draper from JPL, now retired, seeing his early vision unfold in front of us.

I count no less than 20 television cameras, as the press is there to tell the world about what has to be the apotheosis of this incredible exploration. We have all come today to watch the surface of Titan being unveiled, the cloud curtain finally being lifted.

A large screen makes us almost one with the Operations Room, where the signals from space will be coming in, while project leaders are moving in and out of the hall to appear minutes later on the screen and then back again in person, live among us like a movie coming to life. During the presentations in the morning we are getting exiting messages from them, that ground stations on Earth are tracking Huygens *as it falls towards Titan.* "Now at exactly this time the craft is entering the atmosphere at a daunting speed of 21,000 miles [33,600 kilometers] per hour," says the moderator. "That is as *fast as a ride from London to Darmstadt in 15 seconds." Just a while later, Jean-Pierre announces that we are still tracking the craft, which means that* Huygens *should have deployed its parachutes and jettisoned its heat shield. It is all actually happening, and these parachutes did unfold after so many years in their pack.*

And so the day creeps on, bits of information coming to us slowly and building our anticipation and anxiety. In the corridors, pacing around between the briefings, we hear that the Parkes station in Australia continues to hear Huygens *and that it appears to be emitting data, as if the old astronomer himself has taken charge once more. "The craft is alive and should by now have landed."*

At five o'clock we have come back for the press conference, packed together again inside, from scientist to journalist. Al Diaz and David Southwood, the two science leaders from NASA and ESA, are there on the screen. So are Jean-Pierre, and Pascale Sourisse, CEO and spokesperson for Alcatel. I see Mark Dahl and also several of the instrument leaders, and the charming Claudio Sollazzo, in charge of the operations room. Will Huygens *have connected with the mother ship and sent the science data back, as it was supposed to? Will all this actually be as successful as we know it should be, with so much done by so many of the world's top engineers and scientists?*

There it is again! That same tension I saw on television at JPL at the moment of Saturn orbit insertion, and most of all that same tension I felt at the Cape during the launch. But this time I sense this must be the silence of a crowd of millions of people watching inside and outside our room. It is getting stiller and stiller. Claudio keeps bending over to peek at the control panel; it's now 5.17 pm and that's two minutes behind plan. What's happening? Why is it so late? As if two minutes makes any difference over 900 million miles (1.5 billion kilometers). Did the probe data transmission not work after all?

Then it happens. Claudio smiles all of a sudden, and Al and Dave literally fall into each other's arms from joy and emotion – on the screen for all of us to watch and share. Cassini *is talking and starting to tell us about* Huygens' *observations of Titan, some 350 years after the Dutchman did so from our Earth, yet now from no less than the ground of Titan itself! And all of us are cheering yet again like seven years ago, yelling out loud to relieve our tension. Applauding hard to express our genuine feeling of admiration for the women and men before us who followed in the footsteps of these two great astronomers. And this time I myself feel like saying, "Damn right, Dan, it's impressive!"*

Then my mind wanders off and I think of Earle, and how we were discussing this landing many years ago on the deck behind the house. Would it plunge into something very oily? Today he is not here with us in person, yet he is all over this wonder in spirit. I am proud of you, my dear friend, and – more importantly – that is also what your friends of Cassini-Huygens *have been saying all day long.*

EPILOGUE

Now we can relax again, and in European tradition, ESA treats us with drinks and dinner for us to share in the celebrations, joined together – scientists, engineers, press and family – on behalf of the world that watched this mission go all the way. For dessert the team serves us the first picture of Titan's surface. "It looks to us suspiciously close to crème brulée," quips one of the scientists, and I still don't know whether he was kidding.

"Sure looks like a weird place to me," said Dennis as we shook hands warmly on my way out into the cold German night. I keep wondering what planet he was talking about.

In the early parts of this book, we explained the "flow experience," the "whoosh" of excitement when values come together and contrasts fuse. We cited Csikszentmihalyi's proposition that joy, relief, exuberance, and absorption accompany this flow, that the sense of time collapses, that this fusion is often creative, meaningful, and intensely significant. We know and have felt ourselves that this occurs at the moment when value paradoxes are transcended, when seeming opposites become complements.

This does not occur only at the end of proceedings, but continually in so many microcosms. The "Tiger Teams," as the *Huygens* developers called them, had hundreds of such experiences as they completed their projects and passed on their solutions. But it was in Pasadena on June 30, 2004, when *Cassini-Huygens* had to make this very critical maneuver to arrive at Saturn, and then again at Darmstadt in Germany in mid-January, 2005, that all the tigers merged into one incipient surge.

The *Huygens* probe was due to descend on Titan on the morning of the 14th. The late afternoon press conference on the 13th was largely speculation, but early on the 14th the auditorium begins to fill, several thousand people holding their collective breath.

If there are serious doubts in this auditorium, they seem well hidden. Michel Blanc is especially cheerful. He has had his EuroPlaNet proposals to share the science data throughout Europe approved, so that findings from the probe and orbiter will flood the continent, where they could trigger a veritable Renaissance of associated scientific discoveries. Hopefully America will benefit too, and also there has been lots of preparation to receive this bonanza. There is much speculation about the terrain in which the probe will land: soft or solid, wet or dry? Will this all end in a splash, a plop, or a bounce? Alcatel's CEO is assuring everyone that the parachutes will open. Everyone seems to agree, but what are they thinking?

The auditorium is suddenly hushed. Then there is an announcement. A network of radio telescopes across the world is tracking the *Huygens* probe. There is a ragged cheer, because not everyone realizes the implications, but as they inform each other the cheering swells. The probe is in Titan's atmosphere and descending! Then comes a report from JPL in California. It has heard the *Huygens'* chatter, signals as small but as clear as those of a domestic phone call.

Jean-Pierre Lebreton steps up to the microphone. "Looks like we have heard our baby cry." At 60 kilometers above Titan all instruments were turned on. "Now we know they are working." Anne-Marie Schipper announces that the heat shield worked perfectly. The probe has survived the furnace heat of entry.

The press conference ends for the time being, and I (Bram) mingle with the crowd. I encounter Al Diaz of NASA. He has two more paradoxes for us. Titan is a time machine. It reconciles Earth's distant past with its future, its early origins with its destiny as a center of intelligence.

There is a schoolgirl from South Carolina who won an art contest for depicting likely landscapes on Titan. She appears transported to this other world. And Marcello marvels at "the voice of our child" but remembers those who never made it to this celebration, Earle Huckins and Hamid Hassan among others. "Hamid helped to bring us here, so did Earle. This celebration has to be for them."

For David Southwood, head of the ESA science programme, it is also like a newborn child, but one growing up rapidly in the space of seconds. He likens the parachutes opening up to a teenager throwing off his bedclothes and starting to function like a grown-up.

"I am moved, very moved," says Roger Bonnet of ESA, who had approved the project in the 1980s. "It is a great moment. We have discovered a new world. Europe has not done that since the days of Columbus. International cooperation was the key to the quality we all achieved. I want to say to America, 'Let's do this again.'"

Gerard Huttin of Alcatel, the man who led all the *Huygens* work in its development, regrets the world's continued tensions. As a counter-example, the cooperative spirit has prevailed in this case over many difficulties. "It is this sort of venture that gives me optimism," he says.

"We are so very happy," says Anne-Marie, who succeeded Gerard. "I have been with this project only two years, but the spirit of it all is infectious. The indications (of success) are growing stronger all the time. I feel really deeply and just want to go with the flow." She is intelligent, young and attractive – or has everyone in the hall somehow been transformed?

It is David Winfield's birthday. He explains:

> This mission took 16 years of my life. During many of these I thought chiefly about what could go wrong, all those details.... This is the best birthday present I have ever had. Never have I worked on a project with so much spirit. It was very open. It crossed cultures and disciplines. The Tiger Teams that Hamid picked were always multi-disciplinary. For me that was the unique aspect of all this.

Note the implied paradoxes. You worry about *details*, yet enjoy the whole spirit and now the *whole* success. You worry about what could go *wrong*, only to see it all come *right*. David now leads engineering on the International

Space Station, but notes sadly that it lacks the spirit of cooperation manifest here. He must have found the aim of "colonizing space" unappealing.

The press conference reconvened at 5 pm on January 14, 2005. The concern was now beginning to switch from engineering to science. We knew the engineers had made it, but what about the data? Jean-Jacques Dordain, director general of ESA, announced amid cheers and clapping: "The data is coming in!" And it was continuing to do so two hours after touchdown. Most predictions were that the batteries would only last for some 30 minutes amid the permafrost. Dordain was the fourth ESA director since the inception of this mission. He claimed this as ESA's second triumph in a year. The first was *Mars Express*.

Then David Southwood took the microphone. "Now is the time for science," he told us. "The data coming in even as I speak is our gift to posterity. We are present at a historic event that will not be repeated in our lifetimes. It may be the end of the probe, with which we had to lose contact a few minutes ago, but it is the beginning of a science team that will spread knowledge far and wide."

Al Diaz of NASA was next on the rostrum. In the European excitement of the moment it would have been easy to forget that it was NASA and JPL who got the *Huygens* so perfectly to its destination, and it was only fitting for the United States to be represented by a NASA top official. He was still close to tears as he had been the day before, when he wept openly before he finished his words:

> Other missions I have done, well, they tend to come back for longer times. But with *Huygens* all the hard work of all these years had to happen in two hours. It was a very risky mission. I felt completely different this time. When the data came in, it was a very emotional moment and the joy was immense. All we could do was embrace and cry. Something I will remember for the rest of my life. I wished I could have been sitting on the Probe and watch the landscape unfold as if flying over the Alps. It is about more than engineering and science.
>
> Claudio Sollazzo, head of Huygens Spacecraft Operations Unit – ESA

> This has been a long day.... I was really tense when I arrived here this morning, then I felt really relieved when *Huygens* descended after 25 years, over which our lives have been dedicated to this mission. All those relationships, all those partnerships have now culminated in this one moment. It is incredible, and I thank each and every one of you for your contributions, and this tremendous success. There will only be one, first, successful landing on Titan and this is it!

Yet total perfection has eluded the mission. One of the channels on the probe (channel A) was malfunctioning. But redundancies had been built into the system for just such an eventuality, and most science data were safely received through channel B. While the orbiter could not receive this channel, it could be received by radio telescopes around the world, and the missing data from the Doppler wind experiment could be pieced together from their recordings in the weeks and months that follow [as it has].

The press conference now went into many science discoveries. Important as they are, I chose rather to go in search of the principals.

Hans Hoffman, who had done the thermal designs around the *Huygens* and its instruments, told me that all his protected electronics and instruments were working well and had been maintained near room temperature, amid temperature variations that would have shamed "the rings of hell" in Dante's *Divine Comedy*. For him, it was all a technical issue, and his reserve never wavered. But the pride was in his eyes for all to see.

John Zarnecki, from Britain's Open University, was quite frustrated that at this moment of great origination he could find nothing original to say. It was "absolutely fabulous," he told me. "The bringing together of so many characters and actually managing the bloody people was the hard part." Why did it work? The answer is perseverance, a bit of luck, attention to detail, and rigor in the face of unforgiving space. "You had to be already made of stone not to be petrified as today's news came in."

Al Diaz wondered aloud what Earle would have said, had he lived to see this moment. "Something simple probably, like 'Wow!' He was so involved and so excited he'd probably feel like I do now." ESA, NASA, and ASI had attained a stature that none could deny. Perhaps more joint missions would now be possible.

Jean-Jacques Dordain claimed that ESA had already initiated joint missions with China, Japan, Russia, and others across the world. This would add impetus. "We are crowned with success today because we are good! And because we have confidence in ourselves and other nations." I asked him if he wondered whether NASA would dare such a large mission again, as Franco-US relationships were generally not good at this time. "Space can bring us peace and friendship."

Daniel Gautier, whose initial vision made all this possible, seemed more relieved than ecstatic, but his words were significant:

> I have now completed my duty which was, after all, the purpose of my life. I am satisfied that some of the questions we are asking will find answers... I am stupefied that it all worked so well. *Cassini-Huygens* is a beautiful example of unity, of how Europe can unite and what that unity means. I only hope the Americans will join us again in the future for another ambitious mission.

American Bob Brown was his usual insightful self:

> I feel mostly pride, especially also for the Europeans. This is their great spectacular.... The overriding difference on this project was that all

involved truly loved what they were doing. All this diversity among us, national, cultural, disciplinary, has made things more exciting and less difficult. Let's keep doing this! Let us focus on the best parts of what human beings from a world community can accomplish together for the benefit of everyone.

He saw it as the triumph of "professionalism and dedication."

The press conference reconvened on Saturday, January 15. It was high time that the Brits, whose language everyone is kind enough to speak, came up with some stirring words. David Southwood did not disappoint. He was crying too in front of that assembly, from fatigue and more so from emotion, having been up all night to look at the Titan messages. He told his audience:

> We don't have all the science yet. And the gods have reminded us that we're not perfect. One channel failed. But the Titans were somewhere between men and gods and that's us, I guess. Thanks to the radio astronomy of the world, data lost to the winds is being gathered up again.

The audience rallied his strength with yet one more ovation.

Because the engineers had triumphed the day before, Al Diaz, representing the heroes of NASA's JPL, was well rested by Saturday:

> Unlike the scientists, I slept like a baby last night! The level of success that has been achieved in this mission continues to amaze me, and it testifies to the fact that strong interpersonal relationships, extended over long periods of time and which are dedicated to a greater good, can produce incredible results. I am proud of being part of this, and proud that NASA was part of this.

Finally Toby Owen, standing next to Daniel in the coffee break – a symbolism both may not have sensed that moment, but we did – had this to say:

> We had a fantastic global team working on this. I stress again, this is *so* difficult to do – all these different steps that have to go right, and they all do, it's so remarkable, and we are not done yet. There are going to be more extraordinary discoveries. I feel very happy – such a remarkable experience – and we had a lot of fun. There was such a good atmosphere with all these people. It was a lot of work and we enjoyed it. The arguments were sometimes pretty strong but we always got back together. We had a very good mix of people, it worked very well. Investigators never refused to talk to each other, a lot of different possible controversies – engineers versus scientists, Brits and French, and so on – all these combinations and we somehow managed through this very well.

The ideal would be to make planetary exploration a planetary enterprise. Get the Chinese and the Japanese involved, and have everybody together.

So Mihaly Csikszentmihalyi appears to be right. It *can* all end in the experience of *flow*, whether it is flowing words, flowing tears, or the flowing fusion of once contrasting values, joined in a flood of relief and congratulation. So they all got to Saturn and Titan in the end. But do they not also have something to tell us about how better to live our lives, run our teams and organizations? If this is stellar performance, should we not all strive to play our part? There are challenges enough awaiting us.

Appendix A

The science instruments

In some ways, the *Cassini-Huygens* spacecraft has senses better than our own. For example, *Cassini-Huygens* can "see" in wavelengths of light and energy that the human eye cannot. The instruments on the spacecraft can "feel" things about magnetic fields and tiny dust particles that no human hand could detect.

The science instruments can be classified in a way that can be compared with the way human senses operate. Your eyes and ears are "remote sensing" devices because you can receive information from remote objects without being in direct contact with them. Your senses of touch and taste are "direct sensing" devices. Your nose can be construed as either a remote or direct sensing device. You can certainly smell the apple pie across the room without having your nose in direct contact with it, but the molecules carrying the scent do have to make direct contact with your sinuses. *Cassini-Huygens'* instruments can be classified as remote sensing instruments, and fields and particles instruments. These are all designed to record significant data and take a variety of close-up measurements.

The remote sensing instruments on the *Cassini-Huygens* spacecraft can calculate measurements from a great distance. This set includes both optical and microwave sensing instruments, including cameras, spectrometers, radar, and radio.

The fields and particles instruments take "in situ" (on site) direct sensing measurements of the environment around the spacecraft. These instruments measure magnetic fields, mass, electrical charges, and densities of atomic particles. They also measure the quantity and composition of dust particles, the strengths of plasma (electrically charged gas), and radio waves.
Source: Jet Propulsion Laboratory website, 2005

Cassini instruments

Imaging science subsystem (ISS)

The ISS is a remote sensing instrument (think sight) that captures images in visible light, and some in infrared and ultraviolet light. The ISS has a camera that can take a broad, wide-angle picture, and another camera that can record small areas in fine detail. It is anticipated that *Cassini* scientists will be able to use ISS to return hundreds of thousands of images of Saturn and its rings and moons, and possibly discover new moons. (Two new moons have been found so far.)

Team leader: Carolyn C. Porco, Space Science Institute, Boulder, Colo., USA

Radio detection and ranging instrument (RADAR)

RADAR is a remote active and remote passive sensing instrument (think sight and hearing) that will produce maps of Titan's surface and measure the height of surface objects (like mountains and canyons) by bouncing radio signals off of Titan's surface and timing their return. Radio waves can penetrate the thick veil of haze surrounding Titan. In addition to bouncing radio waves, the RADAR instrument will listen for radio waves that Saturn or its moons may be producing.

Team leader: Charles Elachi, Jet Propulsion Laboratory, Pasadena, Calif., USA

Visible and infrared mapping spectrometer (VIMS)

VIMS is a remote sensing instrument (think again human sight) that is actually made up of two cameras in one: one is used to measure visible wavelengths, the other infrared.

VIMS captures images using visible and infrared light to learn more about the composition of moon surfaces, the rings, and the atmospheres of Saturn and Titan. VIMS also observes the sunlight and starlight that passes through the rings to learn more about ring structure. VIMS's role in observing the minor building blocks of life, called "pre-biotic" materials, will be fascinating. VIMS will also study lightning, and throughout the 40 planned orbits around Saturn, the planet's 33 known moons.

Team leader: Robert H. Brown, Lunar and Planetary Laboratory, University of Arizona, Tucson, Az., USA

Composite infrared spectrometer (CIRS)

CIRS is a remote sensing instrument (think again human sight) that measures the infrared light coming from an object (such as an atmosphere or moon surface) to learn more about its temperature and what it is made of.

Throughout the *Cassini-Huygens* mission, CIRS will measure infrared emissions from atmospheres, rings, and surfaces in the vast Saturn system, to determine their composition, temperatures, and thermal properties. It will map the atmosphere of Saturn in three dimensions to determine temperature and pressure profiles with altitude, gas composition, and the distribution of aerosols and clouds. This instrument will also measure thermal characteristics and the composition of satellite surfaces and rings.

Principal investigator: Michael Flasar, NASA/Goddard Space Flight Center, Greenbelt, Md., USA

Ultraviolet imaging spectrograph (UVIS)

UVIS is a remote sensing instrument (think sight) that captures images of the ultraviolet light reflected off an object, such as the clouds of Saturn and/or its rings, to learn more about their structure, chemistry, and composition. Designed to measure ultraviolet light over wavelengths from 55.8 to 190 nanometers, this instrument is also a valuable tool to help determine the composition, distribution, aerosol particle content, and temperatures of their atmospheres. It is capable of taking so many images that it can create movies to show the ways in which this material is moved around by other forces.

Principal investigator: Larry Esposito, University of Colorado, Boulder, Colo., USA

Radio science subsystem (RSS)

RSS is a remote sensing instrument (think again sight, or in this case, hearing) that uses radio antennas on Earth to observe the way radio signals from the spacecraft change as they are sent through objects, such as Titan's atmosphere or Saturn's rings, or even behind the sun. It enables scientists to search for gravitational waves in the universe, and study the atmosphere, rings, and gravity fields of Saturn and its moons by measuring telltale changes in radio waves sent from the spacecraft.

The RSS team also studies the compositions, pressures, and temperatures of atmospheres and ionospheres, radial structure and particle size distribution within rings, body and system masses, and gravitational waves.

Team leader: Arvydas J. Kliore, Jet Propulsion Laboratory, Pasadena, Calif., USA

Radio and plasma wave spectrometer (RPWS)

RPWS is a direct and remote sensing instrument (think all the human senses) that receives and measures the radio signals coming from Saturn, including the radio waves given off by the interaction of the solar wind with Saturn and Titan. The major functions of the RPWS are to measure the electric and magnetic wave fields in the interplanetary medium and planetary magnetospheres, and to monitor and map Saturn's ionosphere, plasma, and lightning from Saturn's atmosphere. The instrument will also determine the electron density and temperature near Titan and in some regions of Saturn's magnetosphere.

RPWS is also adept at determining the dust and meteoroid distributions throughout the Saturn system and between the icy satellites, the rings, and Titan.

Principal investigator: Donald A. Gurnett, Iowa City, Iowa, USA

Ion and neutral mass spectrometer (INMS)

INMS is a direct sensing instrument (think smell) that analyzes charged particles (like protons and heavier ions) and neutral particles (like atoms) near Titan and Saturn to learn more about their extended atmospheres and ionospheres.

INMS is also intended to measure the positive ion and neutral environments of Saturn's icy satellites and rings.

Team leader: J. Hunter Waite, SPRL, University of Michigan, Ann Arbor, Mich., USA

Cosmic dust analyser (CDA)

CDA is a direct sensing instrument (think touch or taste) that measures the size, speed, and direction of tiny dust grains near Saturn. Cosmic dust particles are very small. To understand their size and consistency, they may best be visually compared to cigar smoke – perhaps icy cigar smoke. All of these small particles have an electrical charge within them. Finding out the origins of this cosmic dust, what it is made of, and how it may affect life on Earth has been an ongoing program of research and exploration.

Some of the dust particles are orbiting Saturn, while others may come from other solar systems. *Cassini*'s CDA can also determine trajectories (orbits), as well as the speed of the particles. This allows scientists to determine where the dust originated.

Principal investigator: Ralf Srama, Max Planck Institut für Kernphysik, Heidelberg, Germany

Cassini plasma spectrometer (CAPS)

CAPS is a direct sensing instrument (think touch, taste, smell) that measures the energy and electrical charge of particles such as electrons and protons that the instrument encounters. CAPS will measure the molecules originating from Saturn's ionosphere, and also determine the configuration of Saturn's magnetic field. CAPS will also investigate plasma in these areas, as well as the solar wind within Saturn's magnetosphere.

Principal investigator: David T. Young, Southwest Research Institute, San Antonio, Texas, USA

Dual-technique magnetometer (MAG)

MAG is a direct sensing instrument (think smell or taste) that measures the strength and direction of the magnetic field around Saturn, and its interactions with the solar wind, the rings, and the moons of Saturn. The magnetic fields are generated partly by the intensely hot molten core at Saturn's center. Measuring the magnetic field is one of the ways to probe the core, even though it is far too hot and deep to actually visit.

Magnetometers are instruments that detect and measure the strength of magnetic fields in the vicinity of the spacecraft. MAG's goals are to develop a three-dimensional model of Saturn's magnetosphere, as well as determine the magnetic state of Titan and its atmosphere, and the icy satellites and their role in the magnetosphere of Saturn.

Principal investigator: Michelle Dougherty, Imperial College, University of London, United Kingdom

Magnetospheric imaging instrument (MIMI)

MIMI is a direct and remote sensing instrument (think both sight and smell) that produces images and other data about the particles trapped in Saturn's huge magnetic field, or magnetosphere. This information will be used to study the overall configuration and dynamics of the magnetosphere and its interactions with the solar wind, Saturn's atmosphere, Titan, rings, and icy satellites.

Basically, MIMI studies all the possible sources of energy in and around Saturn, which is crucial to scientists' understanding of the dynamics of Saturn's magnetic field and atmosphere.

Principal investigator: Stamatios M. Krimigis, Johns Hopkins University, Laurel, Md., USA

Huygens instruments

Surface science package (SSP)

SSP contained a number of sensors designed to determine the physical properties of Titan's surface at the point of impact. These sensors also determined whether the surface was solid or liquid. An acoustic sounder, activated during the last 328 feet (100 meters) of the descent, continuously determined the distance to the surface, measuring the rate of descent and the surface roughness (for example, due to waves). During descent, measurements of the speed of sound provided information on atmospheric composition and temperature, and an accelerometer accurately recorded the deceleration profile at impact, providing information on the hardness and structure of the surface. A tilt sensor measured any pendulum motion during the descent and indicated the probe attitude after landing.

Principal investigator: John C. Zarnecki, Open University, Milton Keynes, England

Huygens atmospheric structure instrument package (HASI)

HASI contained a suite of sensors that measured the physical and electrical properties of Titan's atmosphere. Accelerometers measured forces in all three axes as the probe descended through the atmosphere. Since the aerodynamic properties of the probe were already known, it was possible to determine the density of Titan's atmosphere and detect wind gusts. Had the probe landed on a liquid surface, this instrument would have been able to measure the probe motion due to waves. Temperature and pressure sensors also measured the thermal properties of the atmosphere. The permittivity and electromagnetic wave analyzer component measured the electron and ion (that is, positively charged particle) conductivities of the atmosphere, and searched for electromagnetic wave activity. On the surface of Titan, the conductivity and permittivity (the ratio of electric flux density produced to the strength of the electric field producing the flux) of the surface material was measured.

Principal investigator: Marcello Fulchignoni, Paris Observatory, Meudon, France

Gas chromatograph mass spectrometer (GCMS)

GCMS was a versatile gas chemical analyzer that identified and measured chemicals in Titan's atmosphere. It was equipped with samplers that were filled at high altitude for analysis. The mass spectrometer built a model of the molecular masses of each gas, and a more powerful separation of molecular and isotopic species was accomplished by the gas chromatograph. During descent, the GCMS analyzed pyrolysis products (that is, samples altered by heating) passed to it from the aerosol collector pyrolyzer. Finally, the GCMS measured the composition of Titan's surface in the event of a safe landing. This investigation was made possible by heating the GCMS instrument just prior to impact in order to vaporize the surface material upon contact.

Principal investigator: Hasso B. Neimann, NASA Goddard Space Flight Center, Greenbelt, Md., USA

Aerosol collector and pyrolyzer (ACP)

The ACP experiment drew in aerosol particles from the atmosphere through filters, then heated the trapped samples in ovens (the process of pyrolysis) to vaporize volatiles and decompose the complex organic materials. The products were then flushed along a pipe to the GCMS instrument for analysis. Two filters were provided to collect samples at different altitudes.

Principal investigator: Guy M. Israel, Service d'Aeronomie du Centre National de la Recherche Scientifique, Verrieres-le-Buisson, France

Descent imager/spectral radiometer (DISR)

DISR made a range of imaging and spectral observations using several sensors and fields of view. By measuring the upward and downward flow of radiation, the radiation balance (or imbalance) of the thick Titan atmosphere was assessed. Solar sensors measured the light intensity around the Sun caused by the scattering by aerosols in the atmosphere. This permitted the calculation of the size and number density of the suspended particles. Two imagers (one visible, one infrared) observed the surface during the latter stages of the descent, and as the probe slowly rotated, built up a mosaic of pictures around the landing site. There was also a side-view visible imager that obtained a horizontal view of the horizon and the underside of the cloud deck. For spectral measurements of the surface, a lamp switched on shortly before landing and augmented the weak sunlight.

Principal investigator: Martin G. Tomasko, University of Arizona, Tucson, Az., USA

Doppler wind experiment (DWE)

The intent of the DWE experiment was to measure the wind speed during *Huygens*' descent through Titan's atmosphere by observing changes in the carrier frequency of the probe due to the Doppler effect. This measurement could not be done from space because of a configuration problem with one of *Cassini*'s receivers. However, scientists were able to measure the speed of these winds using a global network of radio telescopes.

Principal investigator: Michael K. Bird, University of Bonn, Germany

Sources: Jet Propulsion Laboratory and European Space Agency websites, 2005.

Appendix B
US–European collaboration in space science

1998 Report of the Joint Committee on International Space Programs

Recommendations by the Joint Committee

(Authors' added suggestions in italics, representing their additions to its recommendations)

1. **Scientific support** – the international character of a mission is no guarantee of its realization. The best and most accepted method to establish compelling scientific justification of a mission is peer review by international experts.
2. **Historical foundation** – the success of any international cooperative endeavor is more likely if the partners have a common scientific heritage, that is, a history and basis of cooperation and a context within which a scientific mission fits.
3. **Shared objectives** – shared goals and objectives for international cooperation must go beyond scientists to include the engineers and others involved in a joint mission. *Science objectives need to be set early on.*
4. **Clearly defined responsibilities** – cooperative programs must involve a clear understanding of how the responsibilities of the mission are to be shared among the partners, including well-defined interfaces and efficient communication.
5. **Recognition of the importance of reviews** – periodic monitoring of science goals, mission execution, and the results of data analysis ensure that international missions are both timely and efficient. *Reviews should*

be for the purpose of error correction and mutual respect, not for the purpose of supervision or power.
6 **Sense of partnership** – the success of an international space scientific mission requires that cooperative efforts reinforce and foster mutual respect, confidence, and a sense of partnership among participants. *There is a need for social interaction and funding to do so.*
7 **Beneficial characteristics** – shared benefits such as: unique and complementary capabilities of the partners; contributions of each partner that are considered vital to the mission; cost advantages for each partner; and synergistic effects and cross-fertilization.
8 **Sound plan for data access and distribution** – a well-organized and agreed-upon plan for data calibration, validation, access, and distribution.

The authors would add:

9 ***Leadership*** – *selecting leaders who have demonstrated having the skill to operate effectively in an international, cross-cultural, and cross-disciplinary setting.*
10 **Diversity** – *recognizing the need to construct teams that include specialists from a variety of cultures and technical or scientific disciplines. There may be a need for more specific cross-cultural awareness.*

Source: National Research Council and European Science Foundation (1998).

Appendix C
Text of letter from the director general of ESA to Al Gore, Vice President, United States of America

european space agency
agence spatiale européenne
Paris, 3 June 1994
Jean-Marie Luton
Director General

The Honorable Albert Gore, Jr.
Vice President of the United States
Old Executive Office Building
Washington, DC 20501
USA

Dear Mr. Vice President,
I have recently received a number of disturbing reports that suggest that the continuation of the joint U.S./European CASSINI mission could be

threatened by ongoing Congressional deliberations on NASA's FY95 Appropriations Bill.

I am aware that the House version of the Bill, as marked up by the House VA-HUD and Independent Agencies Subcommittee on June 9, retains the necessary funding for NASA's portion of the mission. However, I am also aware that the House Subcommittee's Senate counterpart is faced with a more stringent budget allocation. I am told that the Subcommittee Chair, Senator Mikulski, has indicated that without an increase in said allocation, termination of a major NASA programme would have to be contemplated, with specific reference being made to the CASSINI mission.

In the field of space science, CASSINI is the most significant planetary mission presently being undertaken by either the European Space Agency (ESA) or NASA, Involving the explication of Saturn, the most complex planet in the solar system and of its Moon, Titan. It is expected to provide at least a ten-fold increase in our knowledge of both bodies as compared to NASA's highly successful Voyager mission.

In making the commitment to participate with the U.S. in 1989, ESA oriented its overall space science programme in order to select this cooperative project, rather than opt for one of a number of purely European alternatives that were proposed at the same time. This decision was taken on the basis of scientific merit and in the belief that the cooperation would be of major benefit to both the U.S. and European scientific communities as well as the international science community in general. Over the past five years, while ESA's Long-Term Space Plan has been forced to undergo a series of significant revisions, driven primarily by our own budget limitations, the Member States have maintained a full commitment to the space science portion of the plan, of which CASSINI is an essential component.

To date, the Member State governments of ESA have committed around $300 Million to our portion of the mission (the HUYGENS Probe that will descend into the atmosphere of Saturn's Moon Titan, and several elements of the Saturn Orbiter Payload), of which two-thirds have already been spent, and have committed to a further expenditure of around $100 Million to see the mission through to completion. These figures do not include the approximately $100 Million contribution of Italy via a NASA/Italian Space Agency bilateral agreement.

The HUYGENS programme has been in the hardware phase for the past four years, with probe delivery to NASA due to take place in two years time. The hardware integration and testing phase started in early May this year.

The CASSINI mission has generated Intense interest in Europe, both within the scientific and engineering community and from the public at large. Approximately 900 European scientists and engineers are working on the programme with more than 30 European institutes and universities involved in the preparation of CASSINI/HUYGENS science.

Europe therefore views any prospect of a unilateral withdrawal from the cooperation on the part of the United States as totally unacceptable. Such an action would call into question the reliability of the U.S. as a partner in any future major scientific and technological cooperation.

I urge the Administration to take all necessary steps to ensure that the U.S. commitment to this important cooperative programme is maintained so that we shall be able to look forward to many more years of fruitful cooperation in the field of space science.
Respectfully,
J. M. Luton

Source: National Research Council and European Science Foundation (1998).

Notes

Introduction

1. See, for example: BBC *Horizon*, October 2004 on "Saturn: Lord of the Rings"; News Release from NASA Headquarters, Washington, 2004-168, 30 June 2004, "Cassini spacecraft arrives at Saturn"; and the Discovery Channel's one-hour special broadcast in August 2004, *Cassini: Rendezvous with the Ringed Planet*.
2. Titan is a moon of Saturn in the sense that it is a satellite of Saturn, yet has many characteristics of a planet. Titan is a very large moon, in fact larger than our sister planets Mercury and Pluto. Hence the term "planet" which we use in this book.
3. Words spoken at the press conference of the *Huygens* landing on Titan by David Southwood, Science Director, ESA.
4. In this book, we prefer the term "paradox" over "dilemma." "Paradox – an apparently self-contradictory statement, the underlying meaning of which is revealed only by careful scrutiny. The purpose of a paradox is to arrest attention and provoke fresh thought [and reconciliation]. Dilemma – from its use in rhetoric the word has come to mean a situation in which each of the alternative courses of action (presented as the only ones open) leads to some unsatisfactory consequence" (*Encyclopaedia Britannica*: Encyclopaedia Britannica Premium Service, 2005).

Chapter 1

1. A most recent example is the way the world reacted to the tsunami disaster in Asia.
2. In the United States, as early as in 1975 the Space Science Board of the National Research Council and its Committee on Planetary and Lunar Exploration had made a general recommendation for an in-depth exploration of the Saturnian system following the *Pioneer* and *Voyager* encounters with the planet.
3. Tobias C. Owen, astronomer, was then at the State University of New York at Stony Brook, and is currently at the University of Hawaii.
4. Daniel Gautier is a scientist at L'Observatoire de Paris-Meudon, France, specializing in Titan aeronomy.
5. Project Cassini: A Proposal to the European Space Agency for a Saturn Orbiter/Titan Probe Mission, November 1982.
6. Wing Ip was a planetary scientist at the time at the Max Planck Institute fur Aeronomie, Lindau, Germany.
7. Giovanni Domenico Cassini (1625–1712), first director of the Paris observatory and discoverer of four Saturn moons and the so-called Cassini division in the rings. He was given the name Giovanni Domenico by his parents, Jacopo Cassini

and Julia Crovesi, after his birth in Italy. However he also used the name Gian Domenico Cassini, and after he moved to France, he changed his name to the French version of Jean-Dominique Cassini. He was the first of the famous Cassini family of astronomers and as such is often known as Cassini I.

8. Technically, this was a Joint Working Group of the National Research Council's Space Science Board and ESF's Space Science Committee. See United States and Western Europe Cooperation in Planetary Exploration (1986).

9. Levy was a member of the Department of Planetary Science, University of Arizona.

10. Scarf belonged to the TRW Space Systems Group, California; Masursky to the Branch of Astrogeology Studies, US Geological Survey, Flagstaff, Arizona; and Fechtig to the Max Planck Institute for Kernphysik, Heidelberg, Germany.

11. The cancelation, in 1981, saved NASA some US$250 million; in fairness to the Americans, they did provide the launch vehicle and use of launch facilities, as well as operations.

12. Wes Huntress later became the NASA associate administrator for space science and applications, and has throughout his career been a strong proponent of this mission and international cooperation. Ron Draper was the *Mariner Mark II* project manager at JPL, and the early *Cassini* project manager. Jean-Pierre Lebreton has been the *Huygens* project scientist since 1990, and currently also leads the overall Huygens project. George Scoon was an engineer at ESTEC who is largely credited for having developed the initial engineering design of the *Huygens* probe.

13. Principal investigators are the scientists who proposed to and were accepted by NASA/ESA to conduct and lead an investigation, including the type of instrument and the science team. Facility team leaders were appointed on instruments already prescribed by NASA/ESA, instruments typically viewed as "standard" for such missions. These instrument teams have team members who all have been selected individually by NASA/ESA based on the merit of their submissions.

14. Hermann Rorschach, 1884–1922 was the Swiss psychiatrist who devised the inkblot test that bears his name and is widely used clinically for diagnosing psychopathology.

15. Congress approved a US$12.2 billion Phase 1 version of the *Freedom* Space Station in early 1987.

16. The Soviets sent four missions to Venus between 1981 and 1983, another two to both Venus and Halley's comet in 1985, and finally the *Phobos 1* and *2* missions to Mars in 1988 each with 22 instruments on board

17. The JPL project manager for CRAF, Ron Draper, remembers proposing this as well (success has many fathers). The *Mariner Mark II* concept was to have one standardized design for multiple missions into the outer solar system – similar in appearance and operations systems, and with similar antennae, electronic controls, and data-processing systems. In this case, one would cruise to a comet/asteroid encounter; one would send *Cassini* to Saturn.

18. Officially authorized by the Senate in October 1990 in the NASA Authorization Act, FY 1991. The Senate had some foresight in those days, as it also ordered the National Space Council to conduct a study on International Cooperation in Planetary Exploration aimed at developing a strategy for cooperation between space-faring nations, with an inventory of space exploration technologies not currently available to the United States, but available from other nations.

19 Part of the "New Start" science program in the late 1980s.
20 In the debate over the Space Station, the "20 percent rule," under which 20 percent of the NASA budget had to be allocated to science, prevailed over those who wanted full funding for the Space Station.
21 Congress kept reducing annual appropriations, causing design changes and delays. This also impacted on ESA's schedules for its contributions.
22 Source: ESA website.
23 Spehalski was *Cassini* program manager for the development phase.
24 Although the betting had been that Clinton would choose his own man after the elections, he let Goldin stay and the latter took up his post formally in April 1993.
25 David Southwood is currently Head of Science at ESA; he was then PI on the Mass Spectrometer instrument, Imperial College of Science and Technology.

Chapter 2

1 His name is pronounced as "chicks-send-me-high" according to one website. See Csikszentmihalyi (1990).

Chapter 3

1 Taichi Ohno of Toyota is credited with being the "father" of lean manufacturing techniques.
2 The Fifth Discipline brings word of "learning organizations," organizations where people continually expand their capacity to create the results they truly desire, where new and expansive patterns of thinking are nurtured, where collective aspiration is set free, and where people are continually learning how to learn together. See Senge (1990).
3 The development phase lasted from 1990 to 1997.
4 *Huygens* project manager – ESTEC, 1990 to 1997. Hamid passed away in 1999 and, like Earle Huckins and JPL's Gary Parker, *Cassini* integration manager, never had a chance to see the arrival at Saturn and Titan.
5 International Traffic in Arms Regulations – ITAR. When the 1998 Military budget bill required that the State Department take over responsibility for its enforcement from the Commerce Department, many space-related products and technologies became subject to even stricter controls. This has greatly impeded international cooperation of the type *Cassini-Huygens* represents. Comments we received from ASI and ESTEC staff were highly critical of this issue, and indicate how difficult it would be under the circumstances to conduct a similarly large, peaceful, international space program again.
6 Gary Parker was deputy associate administrator of the Office of Space Science at NASA HQ, 1996–2001. He died in a plane accident. John Pensinger was JPL's liaison with the Italian space agency ASI and its contractors for this mission.
7 Medea is a character from Greek historical myth, and the title of a play by Euripides, which tells the story of the jealousy and revenge of a woman betrayed by her husband.
8 EuroPlaNet is analogous to the Universities Space Research Association,

incorporated in 1969 and a private non-profit corporation under the auspices of the National Academy of Sciences in the United States. Institutional membership in the Association has grown from 49 colleges and universities when it was founded, to 95 in 2004. All member institutions have graduate programs in space sciences or aerospace engineering. Besides 88 member institutions in the United States, there are two member institutions in Canada, three in Europe, and two in Israel. USRA provides a mechanism through which universities can cooperate with one another, with the government, and with other organizations to further space science and technology, and to promote education in these areas.

9   Chris Jones was then *Cassini* spacecraft development manager, and is currently JPL director, planetary projects.
10 Pygmalion is a character from Greek mythology. He fell in love with Aphrodite, the "ideal of womanhood," and when she rejected him because he was a mortal, he sculpted an ivory model of her. The goddess was flattered and took pity on his unrequited love: she brought the statue to life.
11 Hans Hoffman was leader of the *Huygens* integration and assembly team at Daimler Aerospace, now Astrium, a subsidiary of the European Aeronautic Defense and Space Company NV (EADS).
12 One of the experiments, the Doppler wind experiment, suffered as a result – yet again a back-up solution existed, since Earth's radio telescopes picked up *Huygens*' signals too and it is hoped that this will soon produce the experiment's data anyway.

Chapter 4

1   Buber was a German-Jewish religious philosopher, biblical translator and interpreter, and master of German prose style. His philosophy was centered on the encounter, or dialogue, of man with other beings, particularly exemplified in the relation with other men but ultimately resting on and pointing to the relation with God. This thought reached its fullest dialogical expression in *Ich und Du* (1923; *I and Thou*) (*Encyclopaedia Britannica*: Encyclopaedia Britannica Premium Service, 2005).
2   Principal investigator, VIMS instrument. See Appendix A for a description of all instruments on board *Cassini* and *Huygens*.
3   The "spokes" are mysterious, shadow-like structures in the radial direction of the rings. It is thought that gravitational forces alone cannot account for the spoke structure, and it has been proposed that electrostatic repulsion between ring particles may play a role.
4   There were also three additional interdisciplinary scientists for the Titan investigation with the *Huygens* probe: see Appendix A.

Chapter 5

1   Southwood initially attributed the poem erroneously to Shelley. Symbolically, even Keats' poem had a "hidden defect," just as there was one in the radio

communication between the spacecraft that day. It credits not Balboa, the actual discoverer of the Pacific, but Cortez.
2   This is exemplified in ESA's history, born out of European countries' quest to join forces, and thus increase scale, efficiency, and funding, to develop satellite communications, science programs, and launch vehicles. In fact ESA was formed from two organizations created in the 1960s: the European Launcher Development Organisation (ELDO) and the European Space Research Organisation (ESRO). The advanced technology used by ESA today has had a mandatory science program as a driving force behind it.
3   See Chapter 3 note 5.
4   Hers was "TOST," the Titan orbiter science team. There was also "SOST," the satellite orbiter science team.
5   ESA works heavily through subcontractors, while NASA and JPL design and develop much of the spacecraft in-house. In addition, *Cassini* was a much more complex and larger craft than *Huygens*, and provided "free transportation" to Saturn for many of the European scientists.
6   This was another innovation not tried in earlier JPL missions. See Chapter 6 about the science management and exchange system.
7   See Appendix A.

Chapter 6

1   This is necessary under the US National Environmental Policy Act, 1969. See NASA (1995).
2   Under the Presidential Directive, National Security Council Memorandum no. 25.
3   These RTGs were developed by the US Department of Energy. Radioisotope heater units were also used in both the orbiter and the probe.

Chapter 8

1   NASA's budget grew from US$14.3 billion in 1993 to US$14.9 billion in 2002. These figures are absolute, not restated as current-day equivalents in spending power, so the budget effectively declined considerably in real dollars.
2   The Advisory Committee on the Future of the US Space Program, headed by the Chairman of what was at the time called Martin Marietta, Norman Augustine. The Augustine Committee issued its report in December 1990.
3   *New York Times*, letter to the Editor by Peter J. Gierasch, professor of astronomy at Cornell University, April 2, 1992.
4   Informally, NASA staff would coin the term "being Fisked" in the context of the staff reductions and reorganizations.
5   Goldin also mentioned John Casani in that regard, but the latter was no longer in charge of the mission under his tenure.
6   *Cassini-Huygens* had many interfaces, all carefully defined in memoranda of understanding, particularly those between ESA and NASA, and ASI and NASA. The latter was not signed until well into the 1990s, long after the actual work had started.
7   Coincidentally, this was the year in which ESA's top leader wrote to Al Gore.

Chapter 9

1. The authors agree mostly with the recommendations of the Joint Committee on International Space Programs of the National Research Council and European Science Foundation, made in its 1998 report. In many ways *Cassini-Huygens* meets almost all, if not all of the Committee's guidelines for effective international cooperation in space exploration. The authors have added several suggestions based on our own observations. See Appendix B.
2. See Trompenaars and Hampden-Turner (2001).
3. See Trompenaars and Hampden-Turner (2001).
4. See Watson (1997). What characterized the DNA discovery was that the Nobel Prize was awarded not to the front-line investigators looking through instruments but to the modelers, Francis Crick and James Watson. They knew of the emerging data and created in their minds a replica consistent with these data, a model which later observations confirmed. These events clearly delineate the congruence of creativity and discovery, while the molecule itself revealed a hidden world of connected spirals.
5. The power or act which directs or impels to effort of any kind, whether muscular or psychic (*Webster's Dictionary*).

References and further reading

Arnold, Matthew (1993) *Culture and Anarchy and Other Writings*. Cambridge: Cambridge University Press.
Association of American Universities (2000) Letter to Assistant to the President for Science and Technology Policy, concerning the ITAR Issue, May 16.
Benedict, Ruth (1958) *Patterns of Culture*, Boston, Mass.: Houghton-Mifflin.
Benford, Gregory (1999), *Deep Time*, London: HarperCollins.
Bohm, David (2002) *Creativity*, Boston, Mass.: Beacon Press.
Brandenburger, A. and Nalebuff, B. J. (1996) *Coopetition*, New York: Doubleday.
Broad, William J. (1993) "Scientist at work: Daniel S. Golden, bold remodeler of a drifting agency," *New York Times*, December 21.
Clausen, K. and Deutsch, L. (2001) *Huygens Recovery Task Force: Final Report*, July 27, ESTEC and JPL.
Collins, Jim (2001) "Level 5 leadership: the triumph of humility and fierce resolve," *Harvard Business Review*, January.
Collins, J. C. and Porras, J. I. (1994) *Built to Last*, London: Century.
Columbia Accident Investigation Board (2003) Final Report, August 26, 2003, US Government Printing Office.
Csikszentmihalyi, M. (1990) *Flow: The Psychology of Optimal Experience*, New York: Harper and Row.
De Geus, A. P. (1988) "Planning as learning," *Harvard Business Review*, vol. 66, no. 2, pp. 70–4.
Deming, W. E. (1982) *Quality, Productivity and Competitive Position*, Cambridge, Mass.: MIT Press.
Discovery Channel (2004) *Cassini: Rendezvous with the Ringed Planet*, television program, August.
European Space Agency (ESA) (1998) *The History of the European Space Agency*, Noordwijk, Netherlands: ESA Publications Division.
ESA (2001) *ESA Achievements*, Noordwijk, Netherlands: ESA Publications Division.
ESA (2003) *Italy in Space*, Noordwijk, Netherlands: ESA Publications Division.
Florida, Richard (2002) *The Rise of the Creative Class*, New York: Basic Books.
Frankl, Viktor (1959) *Man's Search of Meaning*, Boston, Mass.: Beacon Press.
Friedensen, V. P. (1999) "Protest space: a study of technology choice, perception of risk, and space exploration," thesis, Virginia Polytechnic Institute and State University.
Friedman, Thomas (2005) *The World is Flat*, London: Allen Lane.
Gibson, William (1962) *The Miracle Worker* (play, later a movie directed by Arthur Penn).
Gierasch, P. J. (1992) Letter to the Editor (by a professor of astronomy at Cornell University), *New York Times*, April 2.

Gleiser, Marcelo (2001) *The Prophet and the Astronomer: A Scientific Journey to the End of Time*, New York: Norton.
Harland, David M., (2000) *Jupiter Odyssey: The Storey of NASA's Galileo Mission*, Chichester, UK: Praxis.
Harland, David M. (2002) *Mission to Saturn: Cassini and the Huygens Probe*, Chichester, UK: Praxis.
Horizon (2004) *Saturn: Lord of the Rings*, BBC television program, October.
Huntress, W. T. Jr. (2003) "The future of human space flight", testimony before Committee on Science, US House of Representatives, October 13.
Huntress. W. T., Moroz, V. I., and Shevalev, I. L. (2003) *Lunar and Planetary Robotic Exploration Missions in the 20th Century*, Space Science Reviews, Netherlands: Kluwer Academic.
Huygens Communications Link Enquiry Board (2000) *Report*, December.
James, William (1949) *Essays in Pragmatism*, New York: Haffner.
Johnson-Frese, Joan (2000) "Dan Goldin's legacy," *Ad Astra*, July/August.
Koestler, Arthur (1964) *The Act of Creation*, London: Penguin.
Kurtin, Owen D. (2005) "Dollars and sense: security and the space business," PBI Media LLC, January 3.
Lambright, W. Henry (2001) *Transforming Government: Dan Golden and the Remaking of NASA*, PricewaterhouseCoopers Endowment for the Business of Government, Syracuse University, March.
Launius, Roger D. and McCurdy, Howard E. (1997) *Spaceflight and the Myth of Presidential Leadership*, Urbana: University of Illinois Press.
Lawrence, P. R. and Lorsch, J. W. (1986) *Organization and Environment: Managing Differentiation and Integration,* rev. edn, Boston, Mass.: Harvard Business School Press.
Lorenz, Ralph and Mitton, Jacqueline (2002), *Lifting Titan's Veil*, Cambridge, UK: Cambridge University Press.
Mair, Victor H. (nd) *Pinyin.com: A Guide to the Writing of Mandarin Chinese in Romanization*, web document by Professor of Chinese Language and Literature, Department of East Asian Languages and Civilizations, University of Pennsylvania.
Maslow, Abraham (1954) *Motivation and Personality*, New York: Harper and Row.
Mintzberg, H. (1994) *The Rise and Fall of Strategic Planning*, New York: Free Press.
Moody, Norman (1999) "Cassini opponents protest craft's August Earth flyby," *Florida Today*, February 28.
NASA (1995a) *Final Environmental Impact Statement for the Cassini Mission*, Solar System Exploration Division, Office of Space Science, NASA (reissued 1997).
NASA (1995b) *Fiscal Year 1996 Budget Estimate*, January, Washington, DC: NASA.
NASA (1997) *Passage to a Ringed World: The Cassini-Huygens Mission to Saturn and Titan*, Washington, DC: NASA.
NASA (2005) *Fiscal Year 1996 Budget Estimate*, NASA, January.
National Research Council and European Science Foundation (1998) Joint Committee on International Space Programs, *US–European Collaboration in Space Science*, Washington, DC: National Academy Press.
Pascale, R. T. (1991) *Managing on the Edge: How the Smartest Companies use Conflict to Stay Ahead,* London: Penguin.
Peters, Thomas J. and Waterman, Robert H. Jr. (1982) *In Search of Excellence: Lessons from America's Best-Run Companies*, New York: Harper and Row.

Rifkin, Jeremy (2004) *The European Dream*, New York: Tarcher/Penguin.
Roland, Alex (1989) "Barnstorming in space: the rise and fall of the romantic era of spaceflight, 1957–1986," in Radford Byerly Jr. (ed.), *Space Policy Reconsidered*, Boulder, Colo.: Westview Press.
Saxenian, Anna Lee (1999) *Silicon Valley's New Immigrant Entrepreneurs*, San Francisco: Policy Institute of California.
Senge, P. M. (1990) *The Fifth Discipline: The Art and Practice of the Learning Organization*, New York: Bantam.
Shapiro, Robert (1999), *Planetary Dreams*, New York: Wiley& Sons.
Siddiqi, Asif A. (2002) *Deep Space Chronicle: A Chronology of Deep Space and Planetary Probes 1958–2000*, Washington, DC: NASA.
Stacey, Ralph (1993) "Strategy as Order Emerging from Chaos," *Long Range Planning*, vol. 26, no. 1, pp. 10–17.
Trompenaars, Fons and Hampden-Turner, Charles (2001) *21 Leaders for the 21st Century*, Chicago: McGraw-Hill.
United States and Western Europe Cooperation in Planetary Exploration (1986) *Report of the Joint Working Group*, Washington, DC: National Academy Press.
Watson, James. B. (1997) *The Double Helix*, London: Penguin Books.
Wheeler, Larry and Siceloff, Steven (2001) "Space Station could cost another $4 billion," *Florida Today*, February 16.

List of contributors

Our gratitude is due to the following members of the *Cassini-Huygens* team, and to others who assisted with the creation of this book:

Kevin Baines
Lori Beth Bandor
Siegfried Bauer
Michel Blanc
Thierry Blancquaert
Scott Bolton
Roger Bonnet
Bob Brown
Irina Bruckner
John Casani
Karen Chan
Leo Cheng
Anne Cijsouw
Kai Clausen
Beverly A. Cook
Angioletta Coradini
Marcello Coradini
John Credland
Patrice Couzin
Mihaly Csikszentmihalyi
Susan Curran
Mark Dahl
Romeo De Vidi
Al Diaz
Ron Draper
Bill Fawcett
Len Fisk
Enrico Flamini
Victoria Friedensen
Marcello Fulchignoni
Daniel Gautier
Tom Gavin
Nicola Gebers
Alberto Gianolio

Martin Gillo
Daniel Goldin
Sherry Groen
Shelley Hampden-Turner
Candy Hansen
Hans Hoffman
Denis Hogan
Brian Huckins
Wes Huntress
Gerard Huttin
Chris Jones
Vesa Kangaslahti
Don Kindt
Charley Kohlhase
Volker Kratzenberg-Annies
Ben Kroese
Ernst Kuliayak
Bill Kurth
Jean-Pierre Lebreton
David Lewis
Martin Liu
John Logsdon
Rosaly Lopes-Gautier
Rosette Lopez
Earl Maize
Dick Malow
Dennis Matson
Con McCarthy
Alfred McEwen
Clare Miller
Bob Mitchell
Robert Nelson
Fred Nordlund
Toby Owen

Bernardo Patti
John Pensinger
Carolyn Porco
David Porter
Christine Prunier
Herve Sainct
Trude Schermer
Anne-Marie Schipper
George Scoon
Pi-Shen Seet
Claudio Sollazzo
Pom Somkabcharti
Roberto Somma
Pascale Soucisse
David Southwood
Daniel Spadoni
Dick Spehalski
Darrell Strobel
Harley Thronson
Fons Trompenaars
Linette Tye
Michel Verdant
Carmen Vetter
Bernhard von Weyhe
Andrew Watson
Julie Webster
Kathryn Weld
Randii Wessen
Gordon Whitcomb
Reed Wilcox
David Winfield
John Zarnecki
Heinz Zimmer
Jim Zimmermann

Index

Note:
References to images (in the central section are prefixed with an I, i.e. I1 refers to image 1. References to notes are prefixed with an n, so 199n1-2 is note 1 to chapter 2, on page 199.

A
Addison, Joseph 159
adjudication versus trading 99
Advanced Micro Devices 164
aerosol collector and pyrolyzer (ACP) 191
Aerospatiale *see* Alcaltel
Aftergood, Steve 117
Agenzia Spaziale Italiana (ASI) 3, 5, I1
Alcaltel (formerly Aerospatiale) 4, 47, 127–8, 130, 178
Alenia Spazio 5, 127–8, 130, 156
announcements of opportunity (AO) 67, 93, 96
antenna *see* high-gain antenna
Apollo 10, 19, 21, 24, 25, 57, 86, 149
Argonauts 67
Ariane rockets 39, 49, 124
Aristotle 35
Arnold, Matthew 166–8
astrobiology 87
atmosphere
 of other planets 9
 of Saturn 8, I9
 of Titan 6, 8–9, 15, 56, I17, 109–11
Augustine Report 149–50, 203n8-2

B
Baines, Kevin 73, 78, 92, 137–8, 141
Bay of Pigs 24, 33
Beagle 39
Beckman, John 27
Benedict, Ruth 175
Benford, Gregory 83
Bird, Michael K. 191
blame, avoidance of placing 41, 58–9, 123, 126, 130
Blanc, Michel 20, 52, 140–1, 179
Blancquaert, Thierry 131
Bohm, David 170
Bolton, Scott 101
Bonnet, Roger 20, 21, 26, 30, 49, 72, 78, 143, 151, 180
Braun, Wernher von 45
Brazilian space missions 39
British United Provident Association (BUPA) 169
Brown, Bob 73, 98, 138–40, 182–3, 186, 202n4-2
Bryan, William Jennings 87
Buber, Martin 71, 202n4-1
budgeting arrangements 94–5
 see also funding, *Cassini-Huygens* cost
Bush, President George (Sr) 27–9, 148–9
Bush, President George W. 70

C
California Institute of Technology 4
cancellation of projects 18, 48, 92, 200n1-11
Cape Canaveral 1, 116, 124
Casani, John 26, 29–30, 51, 87,

93, 97, 98, 100, 138–9, 143, 155, 203n8-5
Cassini, Jean-Dominique 17, 22, 84, 199n1-7
Cassini, Mary 84
Cassini Division 22, 199n1-7
Cassini I15, 111–12, 203n5-5
  design of 4
Cassini-Huygens 107–12
  cost of mission 51
  cuts, real and threatened 9, 29–31, 48, 92–4, 113–14, 149–51, 155–6
  development phase 201n3-3
  Earth fly-by 9, 16, 117
  Jupiter fly-by 16, 88, 101, 107, I2
  Goldin sees as "Battlestar Galactica" 30–1
  instruments *see* instruments
  launch *see* launch
  length of mission 100–2, 107
  memoranda of understanding 203n8-6
  orbits of Saturn 8, 16, 108–9, I2, 132, 178
  orbits of Venus 16, 55
  preparation for launch I3, I4
  project team 4–5
  release of *Huygens* 16 (*see also Huygens*)
  route to Saturn 16, 107, I2
  Spacecraft Systems Office 96, 99
  sponsors and partners I1
  Titan fly-bys 8, 109
  Venus fly-by 107, I2
Cassini plasma spectrometer (CAPS) 188–9
Centaur *see* Titan IV-B
Centre National d'Etudes Spatiale (CNES) 15
certainty–doubt 36
challenge and skill difference 37–8

Challenger disaster 26, 27, 45, 116, 127, 160
  see also Space Shuttle
Chandra X-Ray Observatory 27
chaos–order 38
Cisjouw, Anne 77
Clausen, Kai 47, 58, 70–1, 131
Coalition for Peace and Justice 115–17
cold war 21, 25, 40, 86
Collins, J. C. 38, 169
Columbia 45, 127, 160
comets 18, 112
comet rendezvous asteroid fly-by (CRAF) 26–7, 29, 48, 92
command and control 154–5
commercial secrecy 127–8
communication, human 91, 140–1, 143
communications from *Cassini-Huygens* 177–84
communications technology 5, 102, 125–32
  problems with 125–32
competition–cooperation 40–1, 65–84, 171
complex–elementary 127, 147–57
composite infrared spectrometer (CIRS) 109, 112, 186–7
conflict resolution 7
continuous improvement 47, 50
Cook, Beverly A. 117
co-opetition 68, 79, 82
Copernicus 22
Coradini, Angioletta 89
cosmic dust analyzer (CDA) 108, 112, 188
CRAF *see* comet rendezvous asteroid fly-by
creativity 50, 136, 138, 145–6, 165–6
Crick, Francis 204n9-4
crisis management 41
crisis–opportunity 113–34

Csikszentmihalyi, Mihaly 37–8, 43, 179, 184, 201*n*2-1
cultural issues 58, 76–7, 136–46, 161–6, 175
   cross-cultural misunderstandings 128–30
   culture shaped by leadership 166–9
   see also diversity

D

Dahl, Mark 120, 128, 137, 141–2, 178
DASA/Astrium 4
data
   analysis 87
   collection 88 (*see also* instruments)
de Vidi, Romeo 138
decision making
   level of 64, 78, 135
   methods 78–9, 96–100
   see also trading
Deep Space Network 107
Deming, W. Edwards 39, 47
descent imager and spectral radiometer (DISR) 109, 110, 191
design–review 40, 46, 61
details–whole 180
Diaz, Al 71, 137, 138, 151, 178, 180–3
Dione 22
disaster scenarios 114–21
disasters in space 3, 38, 39
   see also Challenger, Columbia
discovery, history of 22
Discovery program 152
dispute resolution 63, 143
diversity, human/cultural 70–2, 161–6, 194
   see also culture
diversity–unity of views 80
Doppler shift on probe relay 39, 125–32, 156

Doppler wind experiment (DWE) 110, 182, 191, 202*n*3-12
Dordain, Jean-Jacques 181–2
doubt–certainty 36, 46, 119, 171
Dougherty, Michelle 189
Draper, Ron 19, 20, 177, 200*n*1-12, 200*n*1-17
dual-technique magnetometer (MAG) 108, 189

E

EADS 4
Earth 9
Earth Observing System 13, 27, 150
Elachi, Charles 186
elites–equals 41, 77–9, 135–46
emotional intelligence 76, 143–4
Enceladus 109, 112
engineering–science tension 38, 41, 75, 85–105, 143
   and budgets 94–5
   change in emphasis over mission 100–2
   temperamental differences 89–90
   trading system 96–100
engineers in *Cassini–Huygens* mission 19–20, 75, 102, 182–3
environmental impact statement 114–21, 203*n*6-1
Epictetus 113
error correction 39, 50–1, 64, 90, 121–34, 166, 182–3
   and learning 50, 60–1
   reasons for errors 126–30
error–correction paradox 40, 45–64, 114, 171
Esposito, Larry 187
Europa (moon) 107
Europe, ideal of 69–70
   contribution to *Cassini-Huygens* 162
European Science Foundation (ESF) 17

European Space Agency (ESA)
xvii–xviii, 4–5, 17–20, 26–7, 30,
32, 49, 69, 84, I1, 137, 180–1,
203n5-2, 203n5-5
  Bruges meeting 20–1
  Call for Mission Proposals 17
  Darmstadt Operations Centre
    177
  Darmstadt Operations team
    52, 92
  and EU Framework Agreement
    4–5
  European Space Operations
    Centre 4
  funding of missions 23, 32, 92
  joint missions xviii, 182
  lobbying over *Cassini-Huygens*
    31–2, 195–7
  member states 5
  problems of 39
  role in *Cassini-Huygens* 91–2
  Science Programme Committee
    20, 21, 26–7
  Space Science Advisory
    Committee 20
European Space Research and
  Technology Centre (ESTEC) 4,
  47, 69, 88, 125, 137–8
European Union 4
EuroPlaNet 52, 179, 201–2n3-8
expendable launch vehicles (ELVs)
  29

F
"faster, better, cheaper" 30, 148,
  152–4
fault protection 53
  see also error, self-repairing
    features
Fawcett, Bill 54, 98
Fechtig, Hugo 18, 200n1-10
Federation of American Scientists
  117
Fisk, Lennard (Len) 23, 26, 27,
    30, 143, 149, 150, 152, 155,
    203n8-4
Fitzgerald, Scott 154
Flamini, Enrico 52, 60
Flasar, Michael 187
Fletcher, James 27
flexibility, need for operational
  130
Florida, Richard 145–6
Florida
  Coalition for Peace and Justice
    115–17
  spending on space programs in
    25
flow 37, 184
formality–informality 80
France
  contribution to *Cassini-Huygens* 4, 15, 18, 21, 69
  culture 76, 164
  space policy 69
  space program 72
Frankl, Viktor 172
Freedom see International Space
  Station *Freedom*
friendship 15, 143–4, 164–5
  and criticism 46, 57–9, 61
Fulchignoni, Marcello 44, 71, 190
funding
  cost overruns 29
  cuts in 9, 29–31
  for *Cassini-Huygens* 7, 17, 20,
    23, 27
  for space missions generally
    14, 15, 23, 24–5, 27–30, 69,
    92, 148–9
  see also ESA, NASA

G
Gagnon, Bruce 116–17
galaxies, spiral disk 8
Galilei, Galileo 22, 130, 138, 145
Galileo 2, 13, 15, 16, 17, 18, 26,
  39, 78, 87, 95, 96, 102, 141,
  149

Ganymede 17
gas chromatograph and mass spectrometer (GCMS) 110, 190
Gautier, Daniel 15, 17–20, 30, 32, 69, 151, 177, 182, 199n1-4
Gavin, Tom 51, 55–6, 93, 95, 144
generators, radioisotope thermal 4, 116–20, 203n6-3
Germany 164
　contribution to *Cassini-Huygens* 4, 17
Gierasch, Peter J. 203n8-3
Gillo, Martin 164
Giotto 17
goals, superordinate 14
Goddard, Robert 13
Goddard Institute of Space Science Studies 15, 71
Goldin, Daniel (Dan) 14, 30–2, 42, 46, 48–9, 51, 77, 86, 113, 117, 128, 137, 147–57, 160, 201n1-24
Gore, Vice-President Al 31–2, 195–7
Ground System Team 142
Ground System Working Group 142
Gurnett, Donald A. 188

H
Hansen, Candy 91, 100, 101
Hassan, Hamid xviii, 47–8, 56, 58, 76–7, 138, 141, 180, 201n3-4
Hewlett-Packard 171–2
high-gain antenna 55, 107–8, 113, 130, 164
Hipparcos 49
Hoffman, Hans 55, 57, 60, 123, 128, 182, 202n3-11
Hooke, Robert 22
Hubble space telescope 3, 6, 8, 30, 39, 114, 149, 150
Huckins, Earle III v, xviii, 2, 6, 77, 121, 131, 144, 178, 182

Hughes, Michael W. 63
Huntress, Wes 10, 19, 27, 28, 86, 92, 93, 121, 143, 150–2, 154, 200n1-12
Huttin, Gerard 180
Huygens, Christiaan 21, 22, 84, 130
Huygens 92, 109–12
　airflow problem 121–5
　atmospheric structure instrument package (HASI) 190
　descent to/landing on Titan 3, 8, 16, 43, 85, 92, 116, 109–10, 125, 177–80, 199ni-3
　heating units 4, 116–20, 203n6-3
　instruments *see* instruments
　naming of 21
　parachutes 92, 116, 109–10, 178
　release from *Cassini* 16, 115
　see also Cassini-Huygens
hydrocarbons 9

I
Iapetus 22, 108, 112, 131
idealism–reality 40
images from mission xviii, 94, 108
　usage rules 79
　see also images section at center of book
imaging science subsystem (ISS) 94, 108, 186
Incident Surprise Anomalies (ISAs) 123
information sharing 141–3, 170–2
　restrictions on 27, 49, 87, 201n3-5
instruments 74–5, 138–9
　description of individual 108–9, 185–91
　design and construction of 4–5, 124–32
　observation methods 74

protection from temperature
   extremes 55
scanning platform for 30,
   92–4
sharing/sequencing of 21,
   72–3
testing *see* testing
threat of losing 78–9, 92–4,
   98–9, 139
use in Jupiter fly-by 107
use in Saturn orbit 108
see also individual instruments
   by name
International Space Station
   Freedom 18, 19, 21, 26, 27, 31,
   39, 149, 150, 154, 160, 180–1
   cost of 29, 200*n*1-15, 201*n*1-
   20
International Traffic in Arms
   Regulations (ITAR) 87, 201*n*3-5
Io 107
ion and neutral mass-spectrometer
   (INMS) 188
Ip, Wing 17–20, 32, 199*n*1-6
Israel, Guy M. 191
Italy
   contribution to *Cassini-
      Huygens* 5, 21, 31, 32,
      128–30, 164
   communications technology 5
   cultural issues 58, 76, 129,
      138, 165
   space program 72
Italian Space Agency *see* Agenzia
   Spaziale Italiana
ITAR *see* International Traffic in
   Arms Regulations

J
James, William 172
Japan 21, 27
Jet Propulsion Laboratory (JPL) 4,
   15, 19–20, 26, 47, 49, 63–4, 76,
   84, 86, 88, 100, 102, 115, 119,
   131–2, 137, 149, 155, 203*n*5-5

"arrogance" of 20
Spacecraft Systems Office *see*
   Cassini-Huygens SSO
ways of working 30
Johnson, President Lyndon B. 25
Johnson and Johnson 39, 174–5
Joint Board of Enquiry 123
Joint Committee on International
   Space Programs 161, 204*n*9-1
   recommendations of 193–4
Joint Working Group ((NAS/ESF)
   17–18, 142, 200*n*1-8
Jones, Chris 50, 53, 56, 57, 94,
   102, 129, 202*n*3-9
Jupiter 9, 107, I7
   Cassini-Huygens fly-by 16, 88,
      101, 107, I7
   missions to 2, 15–17, 26, 39,
      95
   moons of 16, 107, I7

K
Kaku, Michio 116, 118
Keats, John 85
Keeler, J. E. 22
Kennedy, President John F. 23–5,
   86
Kennedy Space Center 116–17,
   121
Kepler, Johannes 130
Kindt, Don 77, 177
Kliore, Arvydas J. 187
Kohlhase, Charley 54, 84, 96
Kottler, Herb 49, 129
Kroese, Ben 141
Kuliayak, Ernst 123, 125

L
Lambright, W. Henry 155
launch of *Cassini-Huygens* 1, 16,
   31, 64, 77, 89, I5, I6, 119,
   124–5
   risks associated with 114–21,
      124
   threatened delay 121–5

Lawrence, Paul 38
leadership 5, 41–2, 77–9, 135–46, 194
   shaping culture 166–9
   see also elites–equals
learning 34, 72
   accelerated 6–7, 62
   loop 61
   through paradox 174–6
Lebreton, Jean-Pierre 19, 56, 78, 97, 131, 140, 143–4, 177–8, 180, 200*n*1-12
Ledyard, John 98
Levy, Eugene 18, 200*n*1-9
Lewis Research Center (LeRC) 115
life
   origin of 9
   possible in space 19
linearity–circularity 103–4
Lockheed Martin 4
Lomberg, Jon 83
Longfellow, Henry Wadsworth 135
Lopes-Gautier, Rosaly 71
Lorsch, Jay 38
loyalty–dissent 36, 167, 171
Luna 24
Luna 9 24
Luton, Jean-Marie 32, 195–7

M

Magellan 26, 54, 94
magnetosphere 18
magnetospheric imaging instrument (MIMI) 107, 189–90
Maize, Earl 131
management by wandering about 38, 171
Mariner missions 13, 39
Mariner Mark II 17, 20, 27, 200*n*1-17
marker for mission 83–4
Mars 9
   missions to 2, 3, 18, 25, 26, 30, 39, 70, 152–3, 200*n*1-16
   see also individual missions by name
Mars 96 116
Mars Express 39, 181
Mars Observer 30, 39, 54, 96, 98, 116
Mars Pathfinder 152
Mars Surveyor 114
Maslow, Abraham 175
Masursky, Hal xvii, xviii, 18, 200*n*1-10
Matson, Dennis 73–4, 75, 77, 78–9, 92, 94, 96, 97, 98, 137–8, 140–3, 179
Max Planck Institute 17
Maxwell, James Clerk 22
McEwen, Alfred 94
measurement, confusion over units of 45, 114, 127
Medea 51, 201*n*3-7
Metis 16
Mimas 109, 112
Mintzberg, Henry 38
Miracle Worker, The 172–3
missiles 24–5
MIT Lincoln Laboratory 49
Mitchell, Bob 91, 94, 95, 131, 142
Monet, Claude 69
Moon (Earth's), missions to 3
moons
   of Jupiter 16, 107, 17
   of Saturn 16, 22, 18, 112–120, 109–12, 131
   see also individual moons by name
motivation 7, 10–11, 42, 66, 166
   meaning as prime 172–3

N

National Aeronautical and Space Administration (NASA) xviii, 2, 4, 15, 17–32, 46, 48, 76, 79, 87,

92, I1, 111, 117, 137, 139, 147–57, 203n5-5
  Ames Research Center 15
  Deep Space Network 125
  funding of 23, 25, 28, 151, 203n8-1
  Glenn Research Center 4
  Office of Space Science 4
  problems of 39
  Solar System Exploration Committee 15–17
  websites 117
  working methods 57
Near Earth Asteroid Rendezvous (NEAR) 152
Neimann, Hasso B. 190
Nelson, Robert 68
Netherlands, contribution to *Cassini-Huygens* 21
New Millennium program 152
New Start program 201n1-19
Nixon, President Richard M. 25
no blame culture *see* blame, avoidance of placing
nuclear fuel 9, 41, 115
  safety worries 114–21

O
Officine Galileo 5, 129
Ohno, Taichi 47, 201n3-1
optical remote sensing (ORS) 112
orbiter and probe system xvii, 15, 17, 20
  separate funding 20
  see also Cassini-Huygens
organic material in Titan's atmosphere 9, 19
organisation of project 142
Owen, Tobias (Toby) 15, 17–21, 32, 67, 69–70, 74, 177, 183, 199n1-3

P
paradox 2, 9, 33–44, 102–5, 174–6, 199ni-4
  literary titles of 34
  logic of xviii, 7–10, 31–2
  positive paradoxes 39, 79–80
  value paradoxes 34–6, 179
Parker, Gary xviii, 51, 201n3-4, 201n3-6
Parkes Radio Science station 111, 178
Parliament of Scientists 78, 144, 167
particularism 76
Pascale, Richard 38
Paterno, Joe 65
Patti, Bernardo 58, 96
Pensinger, John 51, 201n3-6
Peters, Thomas J. 34, 38, 171
Phoebe 108, I8
Pioneer 10 13, 16, 199n1-2
Pioneer 11 13, 16
planning–adjustment 57
plaque for mission 83–4
play and work 76, 164–5
playing seriously 45–64, 171
plutonium *see* nuclear fuel
Poindexter, John 98
policy, national space
  United States 28–9
  France 69
political aspects of mission xvii, 13–32, 40, 78, 87, 147–57, 160
Polk, Charles 98
Pollack, Jim xviii
Porco, Carolyn 74, 79, 83, 186
Porras, J. I. 38
Porter, David 98, 99
prediction–speculation 103–4
Primavera 63–4
principal investigators xviii, 75, 95, 200n1-13
  see also scientists; individual scientists by name (listed in Appendix A)
problem-solving techniques 57, 59–60, 121–34
Project Science Group 142

protests against mission 116–18
Proton 124
Pygmalion effect 54, 60, 202n3-10

R
radar instrument 109, 186
radio and plasma wave science instrument (RPWS) 108, 187–8
radio science subsystem (RSS) 187
Ragsdale, Lynn 28–9
Reagan, President Ronald 25, 28
RecDel 59, 63–4, 154
redundancies (of equipment) 55–7, 132–3, 181
relaxation (of project teams) 76–7
reviews 193–4
   in building *Cassini-Huygens* 9, 48–51, 127–30
   independent 48, 49, 128
   internal 156
   at NASA 48
   peer review 127
   of policy *see* political aspects of mission; *Cassini-Huygens* cuts, real and threatened
Rhea 22, I13
Riefenstahl, Leni 46
Rifkin, Jeremy 163
risk
   assessment 115–21, 122
   known and ignored 127, 160
   taking 90
robotic missions 13, 24, 26, 86
   see also individual missions by name
rocket technology 100
Rorschach test 23, 200n1-14
Rumsfeld, Donald 98
Russia (and former Soviet Union) space projects 18, 21, 24, 29, 72, 149, 159, 200n1-16

S
safety issues *see* risk; nuclear fuel
Sagan, Carl 21, 31
Saturn 6, 8, 14, 18, I9, I10, 199n1-2
   atmosphere 8, I9
   Cassini-Huygens mission see *Cassini-Huygens*
   distance from Earth 8
   early discoveries 22, 164
   missions to xvii, 16–20, 52
   moons 6, 8, 22, 108, I8, I12–I20, 109–12, 131 (*see also* Dione, Iapetus, Mimas, Phoebe, Rhea, Tethys, Titan)
   rings 6, 8, 22, 74, 108, I9, I10, I11, 202n4-3
   significance to scientists 8, 118
   storms 6, 8, I10
   temperature 8
Savings and Loan crisis 29
scanning platform, removal of 30, 73, 74
Scarf, Fred xviii, 18, 200n1-10
scheduling tools 63–4
Schipper, Anne-Marie 180
Schlesinger, Arthur 24
science
   exchange of information *see* information sharing
   interdisciplinarity 74–6
   as joint discovery 72
   as motivator for *Cassini-Huygens* 13–32, 72, 85, 95–6
   motivator for other space missions 152, 201n1-20
   rivalry with engineering *see* engineering–science tension
   see also instruments, systems science
scientists (on *Cassini-Huygens* project)
   budget for 23, 33
   number of 30
   selection of xviii, 7, 75, 91–2, 93
   see also individual scientists by name

Scoon, George 19, 142, 200*n*1-12
self-repairing features 53, 114
Senge, Peter 47, 201*n*3-2
simplicity–complexity 42, 147–57
simulation–reality 46, 51–3, 127–8
SkyLab 28, 124
Solar Heliospheric Observatory (SOHO) 39
solar wind 18, 107
Sollazzo, Claudio 178, 181
Somma, Roberto 130, 156
Sourisse, Pascale 178
Southwood, David 31, 51, 71, 72, 85, 143, 178, 180–1, 183, 199*ni*-3, 201*n*1-25, 202*n*5-1
Soviet Union *see* Russia
space
    attitudes to exploration xvii, 10, 24, 26, 86–7
    enthusiasm for exploration 25 (*see also* motivation)
    meaning to individuals 23, 172
    race 24, 27, 86–7, 92
space agencies
    purpose 24, 40, 160 (*see also* political aspects of mission)
    see also Agenzia Spaziale Italiana, European Space Agency, NASA
Space Shuttle 3, 19, 26, 28, 39, 46, 86, 113, 124, 150
    cost of 29
    disaster 26, 27, 39
Spehalski, Dick ("Spe") 30, 47, 51, 57–8, 72, 77, 88, 89, 91, 93, 94, 95, 97, 98, 100, 118, 125, 137, 138, 142, 150, 154, 177, 201*n*1-23
Srama, Ralf 188
Star Wars 150
Stop *Cassini!* 115–17
strangers–familiarity 80
    see also friendship, cultural issues

stress 70
Suez Lyonnaise des Eaux 165, 173
Sun, the 9
support–criticism 61, 129–30
    see also friendship
surface science package (SSP) 110–11
SWOT 13
synergy 36, 175
systems science 18, 75, 80, 170
systems, hard and soft 38

T
Tavistock Institute 64
team working 30, 57–8, 65, 73, 141–3, 183
temperatures during mission 55
testing 48–9, 120
    failures in specification for 127–8
    see also reviews
Tethys 22
Texas, spending on space programs in 25
Thatcher, Margaret 66
Thebe 16
Titan xviii, 6, 8–9, 15, 16, I14–I20, 109–12
    atmosphere 6, 8, 15, 56, I17, 109–11
    constituents 19, 110–11
    discovery of 22
    Huygens landing 3, 8, 16, 43, 85, 92, I16, 109–10, 125, 177–80, 199*ni*-3
    importance to scientists 9
    as moon/planet xviii, 199*ni*-2
    surface I18–I20, 110–12
    temperature 16
Titan IV-B/Centaur rocket 1, 4, 27, 53, I5, 113, 124, 137, 149, 156
Titan Orbiter Science Team (TOST) 91, 101, 203*n*5-4
tolerances 55–7

Tomasko, Martin G. 191
Toyota 47
trading system on *Cassini-Huygens* 96–100, 154
Tragedy of the Commons 97
trust, issues of 59–60
  international 92
  and scrutiny 61

U
ultraviolet imaging spectrograph (UVIS) 108, I11, 187
Ulysses 18, 23
United Kingdom, culture 164–5, 167
unity–diversity 36
universalism 76
United States
  contribution to *Cassini-Huygens* 162
  culture of 76–7, 86, 162, 165
  differences from European perspective 162
  meaning of space exploration for 23–5
US
  Air Force 4, 27, 42, 113, 137
  Congress 21, 23, 24, 27, 29, 116, 201n1-21
  Department of Defense 118
  Department of Energy 4, 115, 118–20, 137
  Environmental Protection Agency 118
  House of Representatives Committee on Science 86, 151
  Inter-Agency Nuclear Safety Review Panel 118
  NASA *see* National Aeronautical and Space Administration
  National Academy of Sciences 17
  National Radioastronomy Observatory 109
  National Space Council 200n1-18
  Office of Management and Budget 27
  Planetary Society 84
  Presidency 24, 29, 115, 118 (*see also* individual presidents by name)
  Senate 29, 200n1-18
  White House Office of Science and Technology Policy 118, 150–1
Universities Space Research Association 201–2n3-8

V
values 10, 35–6, 170–2
  added value 35
  value paradoxes 34–5
  see also culture
Venus 9, 55, 107
  missions to 26, 39, 200n1-16
Verdant, Michel 48, 58
vergankelijkheid 2
Vesta 18, 20
Viking 83
visual and infrared mapping spetrometer (VIMS) 89, 108, 109, 138–9, 186
Voyager xvii, xviii, 13, 53, 83, 94
Voyager 1 16
Voyager 2 15, 16

W
Waite, J. Hunter 188
wars
  in Europe 69, 162
  First World 25
  Second World 25
  Vietnam 25
  see also cold war
Waterman, R. H. Jr. 34, 38, 171
Watson, James 204n9-4
weapons, space 24–5
Webb, James 24, 25

Webster, Julie 53–5, 123–5, 131–2
Weld, Kathryn 102
Wessen, Randii 99
Wilcox, Reed E. 63, 115, 117–20
Winfield, David 180

Y
Young, David T. 189

Z
Zarnecki, John 126, 182, 190